SAVING THE PYRAMIDS

SAVING the PYRAMIDS

Twenty-first-century Engineering and Egypt's Ancient Monuments

PETER JAMES

UNIVERSITY OF WALES PRESS

2018

www.uwp.co.uk

British Library CIP Data
A catalogue record for this book is available from the British
Library

ISBN   978-1-78683-250-4
eISBN   978-1-78683-251-1

The right of Peter James to be identified as author of this work
has been asserted in accordance with sections 77 and 79 of the
Copyright, Designs and Patents Act 1988.

Designed and typeset by Chris Bell, cbdesign

Printed by CPI Antony Rowe, Melksham

CONTENTS

LIST OF ILLUSTRATIONS

INTRODUCTION
THE EARLY YEARS

S TANDING FOR THE FIRST TIME in front of the Great Pyramid, one of the world's most celebrated monuments, the subject of study and wonder for centuries, which captured the imagination of great men as diverse as Newton and Napoleon, I never imagined that I would later play a part, however small, in its history.

As I remember my early years in construction and, in particular, when I was deputy building superintendent with the Cardiff City Engineer's Department, dealing with the council's entire building stock, it was barely conceivable that this experience would take me to work on the edge of the Sahara and oversee the repair of the first high-rise stone structure in the world.

The Giza Pyramids of Khufu, Khaefre and Menkaure.

How could I have guessed that resolving the structural problems of pre-Second World War housing on a sprawling estate in the Cardiff suburbs would lead me to own a highly experienced small company specializing in strengthening distressed structures?

The experience of managing a large force of craftsmen engaged in maintaining houses as well as some listed buildings and monuments was invaluable. I was given the privilege, at a relatively early age, of controlling a large workforce of all trades and their workshops. These tradesmen were not judged on their output but by their skills. They were specialists, craftsmen from a period when craftsmanship was measured by its quality.

These craftsmen were so good that they were able to execute the work on the original unfinished relief paintings in Cardiff Castle. The great nineteenth-century architect and designer William Burges, engaged by the Marquess of Bute to restore the castle and build a great house inside its original walls, outlined these paintings while working on the project, but he died young, long before the building programme could be completed, and the reliefs remained merely in outline for more than a century.

This grounding in quality workmanship has been the most important influence in my working life, especially when I started my own company in 1985. We have worked on landmark buildings, including Buckingham Palace, Windsor Castle, Westminster Abbey, York Minster, Lincoln Cathedral, Dover Castle, the White House in Washington and numerous heritage sites worldwide, particularly the pyramids.

My first professional contact with Egyptian buildings, and the love of Egypt which they inspired in me, came in 1994, when I was asked to provide solutions to the problems of strengthening parts of historic Cairo after the devastating earthquake two years earlier. This initial contract was to work with the state-owned Arab Contractors to reinforce the structure of the Al-Ghuri Mosque, which had suffered extensive damage. That was when I was able to visit the Giza plateau to see the pyramids.

The Great Pyramid of Khufu photographed from the Mena House at Giza.

My first glimpse of the Great Pyramid was early one evening, when getting out of my car at a hotel nearby. A pyramid was, of course, what I had come to see, but I was taken aback by its immense scale and size. At that time of the day it was in deep shadow and obliterated almost half the sky.

Like everyone else who visits the monument for the first time, I found myself wondering how it could have been possible for workmen of the day, thousands of years ago, with only their primitive tools, to put up this vast structure – and why on earth the work was undertaken.

My interest, at that time, was merely as a spectator; I felt great respect for the builders but had no sort of professional concern with the method of construction. This changed after I had completed the restoration of some twenty historic buildings in Cairo, when we came into consideration for the structural strengthening of the great pharaonic monuments. The first of these was the Temple of Hibis in the Kharga Oasis, and it was followed by the burial chamber in the Red Pyramid of the Pharaoh Snefru.

Working on these monuments and being in close proximity to them becomes compulsive. It invades the thought processes and develops into an obsession with how and why they were built. I mentioned this to Dr Ghamrawy, a senior consultant engineer with the Supreme Council of Antiquities. He laughed and said, 'That often happens. We call the victims Pyramidiots.'

1 TOOLS OF THE TRADE

W HEN MY COMPANY WAS FORMED in 1985, it was to take exploit patents acquired from a West German engineer. Originally called Cavity Lock Systems Ltd, it had been manufacturing and producing stainless steel cavity wall ties used to overcome the endemic problem in the United Kingdom of corrosion in the mild steel ones.

Wall ties were used from the early nineteenth century to connect single-brick walls together to form a double external wall to prevent moisture passing into the interior of the structure. However, because of inadequate knowledge at that time of corrosion of metals, and particularly of the effect of the black ash mortar used throughout the United Kingdom on bare or galvanized metals, the ties started to corrode, eventually causing the outer skin of the cavity wall to become structurally unstable. The ties could expand to up to seven times their size, forcing the exterior wall to crack and lift, in some cases up to 100 mm, on a traditional semi-detached house. The tell-tale horizontal cracking at every fifth course in the exterior brickwork walls was a sure sign of this failure. Finding a remedy was always going to be difficult. Should the whole tie be removed, or merely bent back into the cavity to protect the external wall from further damage, or would it have to be replaced with a new stainless steel tie?

It was because of the many practical problems of retrofitting replacement wall ties into old structures that I set about looking for a better, more professional solution than the current mechanical expanding stainless steel sleeves or resin injection methods, which caused numerous problems with weak or friable substrates. In my view, the whole tie should be removed and replaced with a new one fitted into both internal and external wall skins.

During my explorations of the industry I came across a Belgian sales engineer looking for a United Kingdom partner for a German inventor, one with experience in using a fabric-socked anchor to solve problems of fixings in the hollow-pot clay blocks used extensively on the Continent. Many of the external walls of West German buildings were constructed in this way, and he had developed some

RIGHT: *The tell-tale horizontal cracking in the external rendering indicating wall-tie corrosion.*

ABOVE: *Corroded wall-ties in an advanced state of corrosion. These are over-cored and removed using diamond drill techniques and replaced with a new stainless steel wall-tie.*

RIGHT: *Cavity wall construction built to prevent moisture entering the dwellings.*

promising ideas using a system that was ideally suited to the problems encountered in repairing or upgrading by adding a new outer wall cladding. Working together, we were able to provide a fabric-socked solution to many of the problems in restoration associated with structural defects, both in hollow-pot and in solid structures.

It was this simple concept that set us off, and we proceeded to develop and fine-tune the idea into a new and more complete way of strengthening structures. After several years of working with these ideas, I decided to acquire the original basic patents and develop the system to encompass a greater range of products and the structures that could be reinforced. It was soon after buying the patents that I decided to create a strong brand name using the current name of the product called Cintec, an acronym for consolidated injection technology.

The first use of the sock technology to overcome the problem of fixing new facades to hollow-pot construction.

The company has grown since then with offices in Canada, America, Australia, India and Egypt. It has won many awards and citations for its achievements over the years, in particular the Queen's Award for Innovation in 2002 for strengthening masonry and stone arch bridges. In 2014, I received the Chartered Institute of Builders Award for the outstanding international project for the year for the work on restoring the ceiling of the burial chamber in the Step Pyramid. This monument is an iconic structure dating from 2700 BC, and it is the earliest high-rise stone building in the world.

Whilst the device we developed looks very simple, it is ingenious in concept: not merely a product but the culmination of engineering know-how and the skills with which we can design a complete system to match the requirements of strengthening individual structures.

DESIGN CONSIDERATIONS

The best way to describe the use of the Cintec anchor is to consider how you would place a reinforcement member into an existing building. Imagine you wanted to provide a reinforced concrete lintel into an existing building. Forget the problem of opening up an area to accommodate the lintel and all the associated support work needed, but merely consider making a wooden box the overall size of the lintel, having calculated the amount of steel reinforcing you require, and place this into the box in the correct position. You then somehow pour concrete into the box until it is full to the top.

Think of our system as a lintel, but instead of the wooden box we have a much smaller fabric tubular sock to replace the wooden box, and accordingly a hole to accommodate the sock is drilled, usually with diamond drilling techniques, into the

Some typical bespoke remedial applications used in stabilisation and repairs.

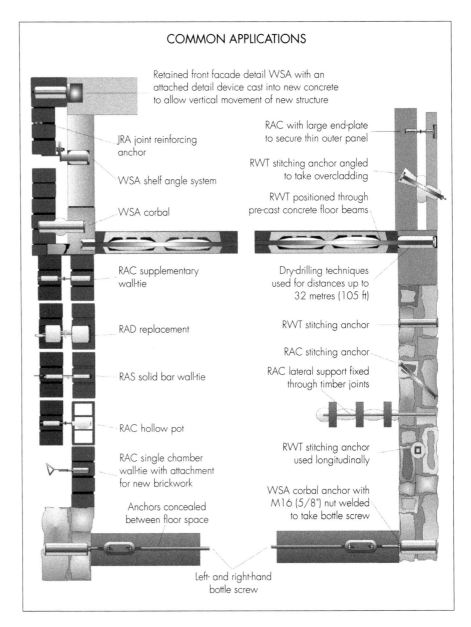

COMMON APPLICATIONS

Retained front facade detail WSA with an attached detail device cast into new concrete to allow vertical movement of new structure

JRA joint reinforcing anchor

WSA shelf angle system

WSA corbal

RAC supplementary wall-tie

RAD replacement

RAS solid bar wall-tie

RAC hollow pot

RAC single chamber wall-tie with attachment for new brickwork

Anchors concealed between floor space

RAC with large end-plate to secure thin outer panel

RWT stitching anchor angled to take overcladding

RWT positioned through pre-cast concrete floor beams

Dry-drilling techniques used for distances up to 32 metres (105 ft)

RWT stitching anchor

RAC stitching anchor

RAC lateral support fixed through timber joints

RWT stitching anchor used longitudinally

WSA corbal anchor with M16 (5/8") nut welded to take bottle screw

Left- and right-hand bottle screw

substrate. Although a large opening is needed when we use a wooden box, we do not have to create a large opening needing struts and props.

Having calculated the main reinforcing bar for the application we are able to provide a complete product that, once we have created the hole, can be made to fit exactly.

The entire assembly is placed in the drilled hole and then a suitable non-shrink grout is pumped from a pressure pot, through injection tubes to the rear of the assembly, filling it from back to front. The grout is placed under pressure and fills the entire assembly. A new development has taken place recently using not only a pressure pot to pump grout into the anchor but also to develop a vacuum at the internal end of the reinforcement. This gives greater control in anchoring the reinforcement, and also a positive feedback to the installer, indicating how the anchor is inflating, which is essential when filling an anchor above the horizontal plane. Once the reinforcement is installed, it remains dormant in the wall until it is required to work structurally.

Whilst the analogy of the lintel mirrors the way the anchor actually works, there are numerous other structural solutions to other, more complex structural problems.

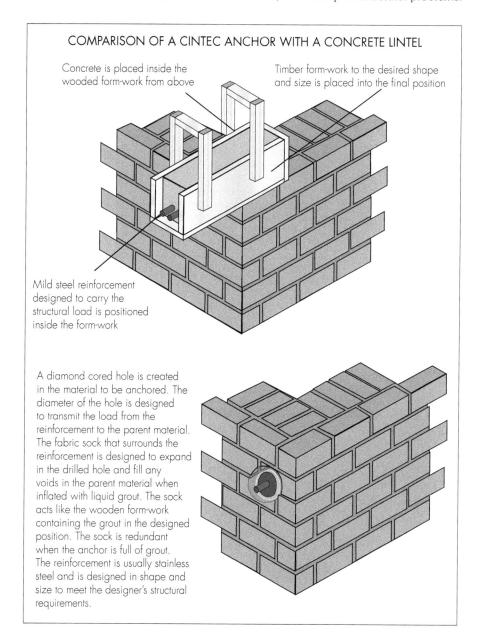

COMPARISON OF A CINTEC ANCHOR WITH A CONCRETE LINTEL

Concrete is placed inside the wooded form-work from above

Timber form-work to the desired shape and size is placed into the final position

Mild steel reinforcement designed to carry the structural load is positioned inside the form-work

A diamond cored hole is created in the material to be anchored. The diameter of the hole is designed to transmit the load from the reinforcement to the parent material. The fabric sock that surrounds the reinforcement is designed to expand in the drilled hole and fill any voids in the parent material when inflated with liquid grout. The sock acts like the wooden form-work containing the grout in the designed position. The sock is redundant when the anchor is full of grout. The reinforcement is usually stainless steel and is designed in shape and size to meet the designer's structural requirements.

An illustration of how the new Cintec vacuum system is used to control grout injection in the anchor assembly. This is achieved by creating a low negative pressure at the extreme end of the assembly as grout is injected into the anchor.

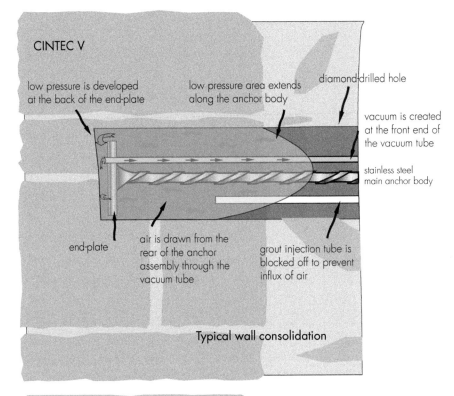

CINTEC V

low pressure is developed at the back of the end-plate

low pressure area extends along the anchor body

diamond-drilled hole

vacuum is created at the front end of the vacuum tube

stainless steel main anchor body

end-plate

air is drawn from the rear of the anchor assembly through the vacuum tube

grout injection tube is blocked off to prevent influx of air

Typical wall consolidation

This shows the advantage of the vacuum system, particularly for installation of vertical positioned anchors.

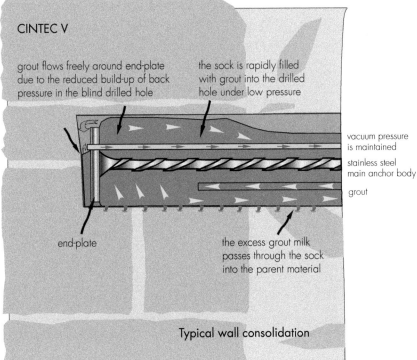

CINTEC V

grout flows freely around end-plate due to the reduced build-up of back pressure in the blind drilled hole

the sock is rapidly filled with grout into the drilled hole under low pressure

vacuum pressure is maintained

stainless steel main anchor body

grout

end-plate

the excess grout milk passes through the sock into the parent material

Typical wall consolidation

Very often conservationists refer to the problem that the reinforcement is too rigid and demand that the grout must be lime-based in keeping with the original structure. Would one take the risk of having a lime-based lintel over one's head? The material and design must be fit for purpose. The original construction was probably built of local materials and needed to be flexible. Structural anchoring and repairs need to be able to mobilize the forces required to lock the structure together. If the designer needs steel reinforcement, this must be connected to the fabric of the structure with a grout that is strong enough to transmit the load to the base material but is contained within the fabric of the sock and does not migrate to other parts of the structure; otherwise the repair will not be effective.

Consider a motor car that needs a steel engine, although you would not want to sit on steel but would prefer leather or fabric. Bearing in mind the problem of compatibility, the designer can find other ways to reduce the extent of the intervention to a minimum by careful use of the smallest possible size of anchor and choice of the most suitable type.

The engineer/designer first calculates the size and profile of the reinforcing member needed, bearing in mind the type and size of load required. All commercially available sizes and profiles are available, such as solid threading bars and grip bars; also from circular, square and rectangular hollow sections and even rolled steel joists.

Normally, because of its high resistance to corrosion and long life expectancy, we use stainless steel graded from 303 or 304 to marine-quality 316 made to the relevant standard. However, we can use mild steel or special memory steel, and even structural grade plastics. We never use galvanized mild steel, as the coating deteriorates within months of being in contact with cementitious grouts.

In order to secure the reinforcement bar in its final position, metal plates made of the same base material as the anchor are fixed to the ends of the anchor to provide a cone of compression locking the anchor body into its final position.

The use of end-plates to create a cone of compression at the ends of the anchor assemblies.

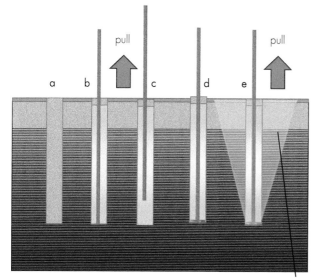

pull pull

a   b   c   d   e

cone of compression

An example of the use of an end-plate

• For example, a circular hole is created in a garden lawn 100 mm in diameter and 500 mm deep (a).

• A broom handle is positioned inside the hole and allowed to stick out of the hole and the hole is filled with sand (b).

• The handle is then pulled and with a little effort is withdrawn from the hole (c).

• If, however, a plate almost the size of the hole is securely attached to the pole and sand is again poured into the hole (d), a cone of compression is achieved, locking the pole in position (e).

The length of the anchor can be varied by the use of connectors to join the anchor sections together in the manner of drain rod connections. The longest anchor we have installed so far is in a cathedral in Newcastle, Australia. This technique was used to make and install a solid 32 mm stainless steel bar 32 m in length. Obviously, in extending the length of the anchor the designer must take into consideration the hole diameter to allow for the size of the anchor body connectors and the extra grout feed tubes, which need a greater internal diameter to allow the grout to flow through the entire length of the assembly. It is also necessary to allow for a larger diameter so that the sock can be placed evenly throughout the length of the drilled hole with an even build-up of grout within the assembly.

ABOVE: *Cintec anchors have been installed up to 32 m in length. This is achieved using connectors and spacers to join and extend the length and size of the assemblies.*

RIGHT: *Illustration of how, by using a larger drilled hole, a greater loading may be achieved in soft and friable substrates.*

COMPARISON BETWEEN A CAR TYRE AND A TRACTOR TYRE OVER SOFT GROUND

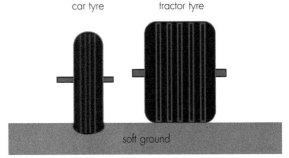

car tyre   tractor tyre

soft ground

Same total load on each tyre
Surface area in contact with the tyre

The larger tractor tyre reduces the overall load by a factor of more than four.

Reinforcing in soft friable material can be achieved the same way by increasing the drilled hole size, thus reducing the load on the parent material to an acceptable level.

reinforcement size remains the same

The hole size must also be increased on long anchors to allow for the additional sock and grout feed tubes.

THE NIKER PROJECT

One of the most exciting new programmes in anchor development has recently completed its design phase after two-year knowledge transfer partnership with Bath University followed by a commission from Niker 7, a new integrated European Union initiative for cooperation between partners from universities and research centres to develop protection with minimum intervention for heritage sites against earthquake damage.

Mechanisms to prevent structural failure, construction types and materials, intervention and assessment techniques will be cross-correlated to develop new methodologies. Traditional materials will be complemented and enhanced by innovative industrial processes, and new high-performance elements will be developed. Advanced numerical studies will allow the parametrization of the results to optimize simple design procedures. In addition, advance monitoring and early-warning techniques for intelligent interventions will be developed.

Earthquake destruction accounts for a substantial part of the global heritage losses caused by natural hazards, as well as very high loss of human life. The post-earthquake survey of damage makes possible scientific understanding of the drawbacks and limitations of new technologies and approaches applied in earlier repair works.

Working in real-application conditions, the project aims to develop and validate complete and diversified innovative technologies and tools for the systematic improvement of their ability to withstand the stress of seismic shifts.

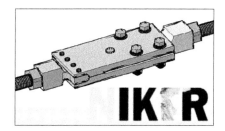

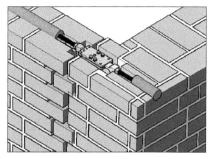

TOP & CENTRE: *Testing at Bath University for project Niker 7.*
BOTTOM: *Pull-out tests at Bath University.*

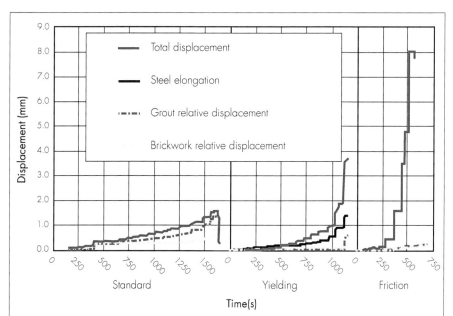

LEFT: *Graph of tests loads.*

A three-year programme began in January 2010, with the active participation of eighteen universities, research centres, public organizations and companies from twelve countries in Europe and the Mediterranean basin. The coordinator is the Department of Structural and Transportation Engineering of the University of Padua, Italy, and the participants were:

The University of Padua.
The Federal Institute of Materials Research and Testing (Germany).
The Institute of Theoretical and Applied Mechanics (Czech Republic).
The National Technical University of Athens (Greece).
The Politecnico of Milan (Italy).
The University of Minho (Portugal).
The Universitat Politècnica de Catalunya (Spain).
The University of Bath (UK).
The Gazi University (Turkey).
L'Ecole Nationale d'Architecture (Morocco).
Cairo University (Egypt).
The Israel Antiques Authority (Israel).
Bozza Legnami (Italy).
Cintec International (UK).
Interprojekt (Bosnia and Herzegovina).
S & B Industrial Minerals (Greece).
ZRS Ziegert Seiler Ingenieure (Germany).
Monumenta (Portugal).

A further development of an energy absorbing dissipative seismic device.

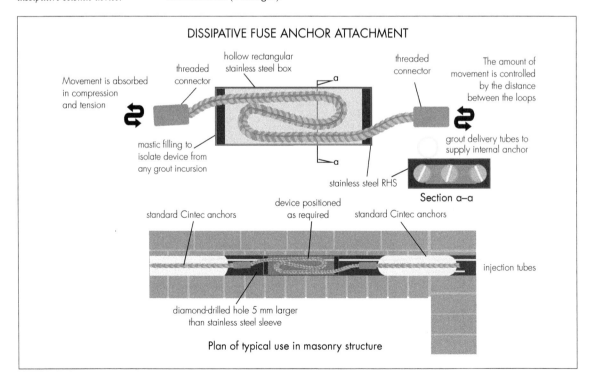

DISSIPATIVE FUSE ANCHOR ATTACHMENT

Movement is absorbed in compression and tension

threaded connector

hollow rectangular stainless steel box

threaded connector

The amount of movement is controlled by the distance between the loops

mastic filling to isolate device from any grout incursion

grout delivery tubes to supply internal anchor

stainless steel RHS

Section a–a

standard Cintec anchors

device positioned as required

standard Cintec anchors

injection tubes

diamond-drilled hole 5 mm larger than stainless steel sleeve

Plan of typical use in masonry structure

The first concept of a dissipative method developed for Niker 7 was very expensive and its use could only be considered on a structure that was especially noteworthy and of great historical value. Consequently, the project is still ongoing for Cintec with the aim of using a similar principle but with lower production costs and greater ease of installation in smaller-diameter drilled holes. This would bring it within the range of the vast number of endangered domestic dwellings in seismic regions. The product has been called a dissipative fuse.

THE PARENT MATERIAL

The strength of the parent material (the base material to be strengthened) and the mortar can govern the anchor capacity. Design checks on its capacity can be based on the resistant strength, according to national standards of the construction, to the anchor force. When the parent material or mortar strength is not determinable, the capacity of the material can be worked out by means of local site tests.

In general, this means that the greater the strength of the parent material, the smaller the drilled hole needed. When dealing with weak and friable parent material, it is necessary to increase the size of the drilled hole to spread the load over a greater area – exactly as a large tractor tyre enables movement over soft and marshy terrain. This allows the anchor to be used in a large variety of substrates.

THE DRILL HOLE

Placing the reinforcement correctly is probably the hardest part of the intervention and can cause most problems for the installer. Positioning the anchor assembly is vital for the designer. As the hole is drilled, any variation in the quality and type of the parent material causes a major problem for the drilling technician. If the substrate is too hard it will require a different compound for the cutting head. If it is soft and friable, and collapses when the core section is removed, or if there is a combination of all these factors, that is the worst-case scenario.

The designer calculates the correct diameter to be drilled, which will depend on the bond strength between the grout and the anchor body, and the grout between the sock and the substrate. Where there is a problem in defining the strengths, the recommendation is to undertake pull-out trials to assess the bond strength of the anchor.

When dealing with historic structures, the impact of drilling must be reduced to the minimum. Mostly the techniques we use are based on diamond drilling technology, particularly, dry-diamond drilling cooled by moist air – what we call keyhole surgery. However, we have used every type of drilling technique, from rotary percussion drills to large hydraulically powered equipment and power mining barrels in a great variety of projects.

Another technique that we have recently patented is the ability to create grooves in the drilled hole to increase the grout-sock bond capacity, particularly when using short anchors in high-strength material that requires large pull-out capacity.

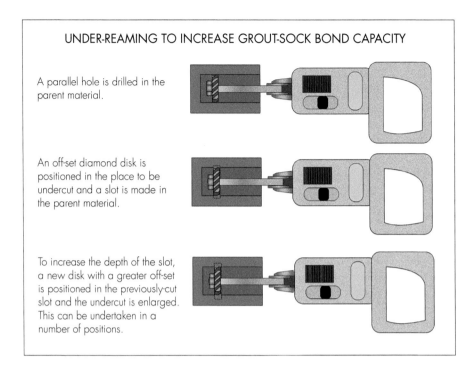

UNDER-REAMING TO INCREASE GROUT-SOCK BOND CAPACITY

A parallel hole is drilled in the parent material.

An off-set diamond disk is positioned in the place to be undercut and a slot is made in the parent material.

To increase the depth of the slot, a new disk with a greater off-set is positioned in the previously-cut slot and the undercut is enlarged. This can be undertaken in a number of positions.

THE GROUT

Presstec is the product name for the cementitious grout. It is a factory-produced mix with graded aggregates and other constituents which, when combined with water, produce a pumpable grout that exhibits good strength with no shrinkage and no added undesirable elements or resins. It is made in accordance with German Din standards and is checked in the factory laboratory both during manufacture and at final dispatch, and also periodically by a recognized external testing house.

Presstec can also be varied in its manufacture to match the substrates and other conditions such as fast-setting grouts, a rapid-hardening grout, several sulphate-resisting and lime-enhanced grouts.

THE SOCK

The fabric sleeve is a specially woven polyester-based tubular sock with expansion properties to match the diameter of the drilled hole and substrate. The mesh is designed to contain the aggregates of the mixed grout while still allowing the cement-enriched water (milk) to pass through the sock, both sizing and bonding the grout to the substrate. The sock is manufactured in sizes from 20 mm to 300 mm in diameter and is adjusted to suit each individual application. It can also be made of different materials to produce a more resistant interface needed in special circumstances.

The sock is made to our exact specification. The quality of the thread and hole mesh sizes are made to correspond to the anchor type, diameter and overall length. Every anchor sock is measured and cut on a machine designed to control the amount of sock taken from the roll without subjecting it to longitudinal tension. The result is that the manifested product has the correct length and diameter of sock needed for each anchor size without compromising a parallel expansion of the sock when inflated.

The construction of the anchor assembly can vary considerably to provide the designer with an all-encompassing solution to the design problems. Some parts can be socked and others left uncovered. This is done by locating grout feed tubes in the sections to be grouted whilst areas that are not to be inflated can be left without a socked attachment. Some anchors can have detachable grout tubes that can be removed during installation if required, whilst multi-bar anchors can have the feeding tubes in their core,

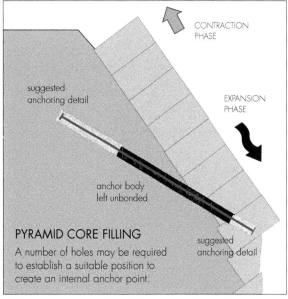

suggested anchoring detail

CONTRACTION PHASE

EXPANSION PHASE

anchor body left unbonded

PYRAMID CORE FILLING
A number of holes may be required to establish a suitable position to create an internal anchor point.

suggested anchoring detail

Smart anchors to be installed at the point of failure in the Bent Pyramid.

and these tubes can be removed and withdrawn as the anchor is inflated.

After careful unwrapping of the anchor from the factory delivery tube, the entire assembly is accurately positioned inside the drilled hole, which must be inspected, and any dust or debris removed, prior to the installation of the anchor to ensure that the sock is not snagged on any obstructions.

The introduction of grout into the assembly is undertaken using an air-driven pressure pot capable of moving the pre-mixed liquid grout at 3 bars of pressure. The grout is pumped to the rear of the assembly and passes from back to front so that the operator knows when the assembly is full. This process, known as Cintec V, has recently been improved with more control and better results. Besides pushing the grout into the assembly, we also pull it by using a low-pressure vacuum created at the very rear of the anchor, thus reducing the back pressure in the drilled hole. The vacuum is created by a Ventura device connected in parallel with the injection tube, thus increasing the speed of injection, removing any air trapped in the sock during the inflation process, and providing the operator with a more positive indication that the assembly is inflated. This is particularly important when inflating and installing anchors at an angle above the horizontal.

Besides testing on individual sites when projects are in progress or the substrate is being assessed, the main elements of the product have been tested by international laboratories such as the Building Research Establishment, the Transport Research Laboratory, and the European Testing Laboratory as well as many universities worldwide. The tests include the accelerated-ages test for forty years without any deterioration of the product, a two-hour fire test, freeze-thaw tests for working in cold climates together with pull-out, impact and seismic tests on a shaking table.

2 RESTORING HISTORIC CAIRO

O N 12 OCTOBER 1992, with its epicentre near Dahshur 35 km south of Cairo, an earthquake measuring 5.8 on the Richter scale struck the Egyptian capital. Although relatively small, this seismic shock was unusually destructive, causing 545 deaths, injuring 6,512 people and making 50,000 homeless. It was the most damaging recorded event ever to affect Cairo.

Most casualties resulted from people being trampled in the rush to get out of shaking buildings. Some 350 structures collapsed, mainly old and unreinforced masonry buildings, and 8,000 were damaged. An estimated 3,500 adobe-type buildings in the surrounding villages were razed to the ground. One of the worst disasters was the collapse of a fourteen-storey non-ductile concrete building in the city, causing sixty-one deaths. Unfortunately, there were no motion-sensing and measurement instruments in the region.

Egypt did not have a seismic design code before the earthquake, and most buildings were not designed for earthquake loads, though it was later noted that, during the strengthening repairs of the historic monuments, the ancient builders had used palm trees hidden inside the walls. These were laid in horizontal lines at 1.2-metre heights in the centre of the walls to give them seismic resistance by adding ductility to the walls.

Richard Swift, a structural engineer with Gifford and Partners, who had worked on the restoration of Windsor Castle after the fire, had been seconded by UNESCO to examine the damage done to the Great Pyramid at Giza. Graham Fletcher of English Heritage, who had used our products on heritage structures, had advised him to contact John Dimmick, our sales manager, to discuss the possibilities of using the firm's product in Egypt. He also told me of the appalling condition of the historic buildings in Cairo and advised me to send Dimmick out to see the problems at first hand and to find a good local agent to represent us.

Swift knew of an agent who also represented a United Kingdom company in the supply of stainless steel fixing and introduced Dimmick to them. The company, Intro Trading, had an office a short distance from Cairo's Tahir Square, and

The damaged interface between the mosque and the minaret.

through its owner, Mamdouh Abbas, and his contacts with Arab Contractors and the Supreme Council of Antiquities, we were invited to make recommendations for the repair of a mosque known as Al-Ghuri.

The Madrasa and Kangah of Sultan Al-Ghuri is monument number 189 of the Muslim monuments under the care of the Egyptian Antiquities Organization. It dates back to 1503–4, when it was built in memory of the Sultan, who had been killed in battle shortly before. The funerary complex is situated in the Fahhamin quarter of Cairo in Al Muizz Street. On the west side of the street is a madrasa mosque, with a four-storey rectangular minaret approximately 50 m high, and on the eastern side the mausoleum, as well as the Sabil Kuttab (fountain and Quranic school) of Abd Al-Rahman.

The inspection revealed serious, long-standing problems:

The floor of the mosque undulates dramatically, evidence of a very significant foundation problem of the masonry vaults supporting the floor. Attempts have been made in the past to underpin the sleeper walls supporting the vaults. These have failed. All the walls of the Mosque exhibit very severe fractures. These are a function of the problems brought about by the rising contaminated ground water table as well as the shaking due to the earthquake. Further problems in the external walls are being caused by the activities of the shopkeepers trying to enlarge the space available for selling their wares. As a consequence, sections of the masonry have been demolished at ground floor level to create this additional space.

TOP: *Displaced voussoirs on an internal arch.*
BOTTOM: *Propping to support a decorative lintel.*

The Mosque of Al-Ghuri was in a very delicate state of equilibrium. Despite having survived for nearly 500 years, the toll of the rising water table, earthquakes and neglect had brought its deterioration to the point of collapse. Urgent measures were required to introduce some structural strength and stiffness into the building.

We had been told that an Italian company using a micro piling system had underpinned the Madrasa in the past. It was therefore necessary to tie elements of the superstructure together. The problem was that the external walls are extremely high laterally and very vulnerable to forces of another seismic event. As with most ancient structures they are built of two skins, a good outer face and a good inner face with rubble core filling making up the bulk of the structure.

It was clear that our stitching system should be used throughout the mosque. Some of the stitches would need to be as long as 12 m and would serve to stiffen each individual wall as well as the external walls to resist seismic influence on the mosque.

The large arched openings in the mosque were a particular cause for concern as points of weakness in the structure. Longitudinal ties in each of the stone faces of the wall above the arch would serve to resist the thrusts naturally produced by the arch as well as seismic loading. Stitching anchors would also be used

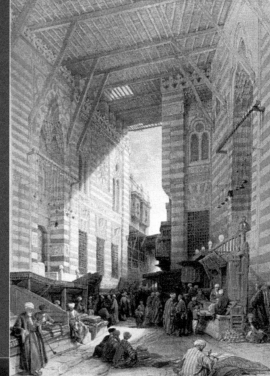

ABOVE: *External view of Al-Ghuri Mosque from the main road.*
RIGHT: The Silk Market, *painting by David Roberts.*
BELOW: *The Madrasa and Kangah of the Sultan Al-Ghuri.*

Temporary timber support to a damaged arch.

Detail of defective roof opening.

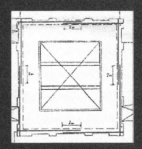

ABOVE: *Placing anchors in position prior to grouting.*

ABOVE: *Roof upstand after anchoring and making good.*
LEFT: *Illustration of the remedial anchor positions.*

to tie the roof structure to the perimeter walls and create diaphragm action. Again, this was an internationally recognized method of introducing greater stiffness and earthquake resistance into a structure. The advantage of our anchor system is that it contains the grout to be used within a fabric tubular sleeve, which controls the grout quantity and flow. Its impact upon the existing structure is neutral and it lies dormant until its strengthening effect is activated by structural forces.

Gifford and Partners, a structural engineering company with a fine reputation, were appointed to conduct a mathematical numerical analysis of the structure using a finite-element technique on the mosque. This provides a graphical animation of seismic events giving a model of how the building reacts in an earthquake in almost real time. As part of a general programme of proposed repairs to this mosque and others affected by the earthquake, it was agreed that this computer simulation would be made to substantiate the proposed method of repair, both to make good the damage caused by the 1992 earthquake and to provide greater resistance in future earthquakes.

ABOVE: *Marking out an anchor position in a heavily decorated internal wall of the mosque* (TOP). *Drilling through and pinning damaged voussoir* (BOTTOM).

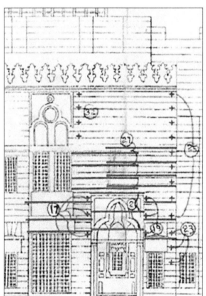

ABOVE: *Position of the anchors on an external corner detail.*
TOP RIGHT: *Position of the anchors on the defective lintel.*
BOTTOM RIGHT: *Position of the anchors in the main mosque arch.*

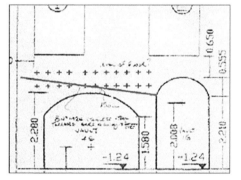

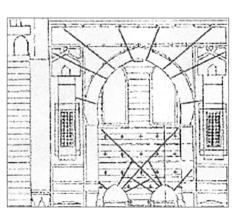

The cracking pattern caused by the earthquake. The white lime patches are tell-tales put on the cracks by the Council of Antiquities after the seismic event. These would have been dated and would have cracked across if further movement had taken place.

One of the particular problems facing the engineers was the large vertical crack that ran the entire length between the minaret and the main body of the mosque. A decision had to be made whether to stitch across the crack and reconnect the minaret back to the main body of the mosque. The simulation indicated that the crack should not be rigidly tied because the vibration frequencies of the two parts were completely different and causing a detrimental pounding action between the two structures. We concluded that if the minaret were tied to the mosque it would suffer major damage from compressive forces as it knocked against the mosque. However, if an expansion joint were provided, the vibration forces between mosque and minaret would be contained. The modelling also confirmed the proposed general arrangement and location of stitching anchors in the adjacent areas.

Further confirmation came when we obtained the latest state-of-the-art discrete-element simulation animation software that we were starting to use on masonry bridges to assess their strengths. This was later shown to be correct when the software developed by Rockfield Software in Swansea was verified in a number of tests at the Transport Research Laboratory in which full-scale bridges were tested to destruction. The software was verified and shown to be accurate to +/– 15 per cent following the full-scale tests on two masonry arch bridges in 1999 and 2001.

This would be the first contract we had in Cairo, and after much debate in the Supreme Council of Antiquities and a number of demonstration tests in various mosques the contract to repair Al-Ghuri Mosque was signed, and the work commenced in May 1998, and was completed approximately a year later.

It was soon after the completion of this project and the acceptance of our skills and technology, that we were invited to tender for a much larger project, the reconstruction of historic Cairo, which suffered the greatest damage following the earthquake. From Bulaq and southwards along the Nile as far as Gera, on the west bank, 350 buildings were completely destroyed and 9,000 others severely damaged. Some 216 mosques and maqaads (courtyards) were badly damaged, particularly the older masonry and adobe structures. The effect was aggravated by the liquefaction of the soil (the soil changing into a liquid mud) reported at the epicentre and the old historic quarter of the city causing the buildings to tilt and collapse as they settled in the soft unsupportive ground.

A tender to reconstruct the greater part of old Cairo was prepared. A number of general contractors were awarded contracts to undertake the demolition of the damaged structures, and to restore them to their original condition. Cintec was invited to tender for the specialist anchoring and reinforcing of the damaged structures in order to save as many of the historic buildings as possible. Initially the contract was for seven of these structures, and the original contract was extended, on completion of the initial tender, to include more mosques.

Richard Swift was requested carefully to examine each building and prepare a structural analysis of the work needed to restore and to strengthen the buildings against future earthquakes. The first four buildings were surveyed and analysed to restore their structures and the drawings were given to Rockfield Software to prepare a mathematical model using their discrete-element techniques to analyse the behaviour of the structures during future seismic events. This programme was

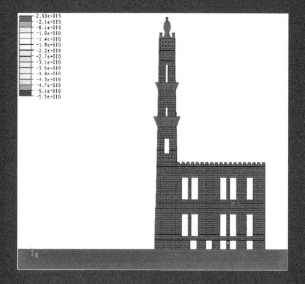

ABOVE LEFT: *Animation used to predict the behaviour of the mosque and minaret during a seismic event.*

ABOVE RIGHT: *Animation of the behaviour of a mosque in seismic conditions.*

LEFT: *Animation of the amount of internal cracking created by the 1992 earthquake.*

BELOW LEFT AND RIGHT: *Animation of a mosque strengthened and subject to a seismic acceleration in a 'what if' scenario.*

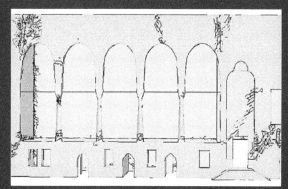

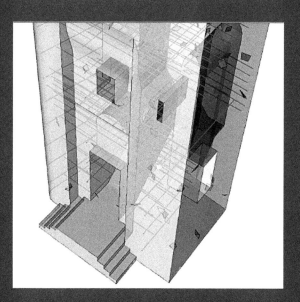

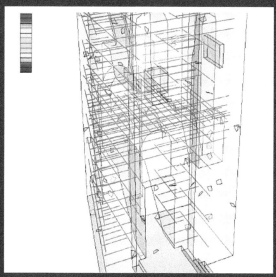

Timber support work to the facade of a maqaad prior to the commencement of work.

also able to provide a graphical moving three-dimensional view of the existing structure and the resultant damage. It was able to check and position the anchor reinforcement mathematically for its most effective position and confirm its suitability for mitigating the damage of future earthquakes. Throughout all this, we worked under the supervision of the Egyptian Antiquities Organization together with five major contractors.

The installation work was carried out in the middle of very densely populated and confined areas. The everyday activities of buying, selling and haggling were carried on apace by the local inhabitants, regardless of the cacophony of the general construction activities, causing noise and confusion reminiscent of Dante's Inferno.

An amusing incident occurred when I was in the middle of discussions with the main contractor, Arab Contractors, at a project in Cairo. I was with Ayman Abbas, eldest son of our agent, who had just graduated from the American Egyptian University of Cairo and was cutting his teeth on the contractual construction implications of restoration work. We had had a meeting on site at the mosque of the Sultan Al-Ghuri in the middle of the old quarter in Cairo and were returning to the head office of Intro Trading, travelling in the company's chauffeur-driven Mercedes 320 saloon car, of which there were not many in Cairo in 1998.

The traffic in those early times was even more chaotic than nowadays. No one took any notice of traffic lights and junctions were policed by soldiers with rifles and loud whistles. As we travelled through the centre of the city, Ayman, who was sitting in the front of the car next to the chauffeur, directed him to take a slip road leading to an overpass near the Cairo Museum. Unfortunately, this was guarded by a soldier plus his rifle, who immediately stopped the car and engaged in a heated debate in Arabic with the driver. The result was stalemate, which was only resolved when Ayman waved the soldier to his window and removed his dark glasses, put them into the top pocket of his immaculately tailored suit and started berating the soldier. The soldier immediately stood to attention, stopped all the traffic and waved the car through, presenting arms at the same time. After a while curiosity got the better of me and I asked the burning question, 'What did you say to him?' Smiling, Ayman said, 'I told him you were the Italian ambassador.'

The structures to be restored included:

Al-Ghuri Mosque
As Silahdar Mosque
Mahmoud Moharam Mosque
Bestaq Maqaad
As-Sarghitmish Mosque
Pabers Mosque
Maqaad Mamy As Saifi
Maqaad Waqf Al Mulla House
Maqaad Palace of Emie Tas
Maqaad Ash Shabshiri House
Maqaad Qayt Bay House

ABOVE: *Bracing to secure walls out of alignment.*
LEFT: *Temporary timber support work to defective arches.*
BELOW: *Typical evidence of seismic cracking.*

RIGHT: *Raking shoring to support walls out of alignment.*
BELOW: *Front elevation of a maqaad before work commenced.*

LEFT: *Severe damage at the As Silahdar Mosque.*

ABOVE LEFT: *Installing an anchor in a heavily built-up area before grouting in position.*

ABOVE RIGHT: *Working in difficult conditions.*

LEFT: *Installing anchors at roof level.*

ABOVE LEFT: *Diamond drilling in a door opening in a congested area.*
ABOVE RIGHT: *Carefully installing an anchor.*
OPPOSITE: *Interior of Al-Ghuri Mosque.*

In total over 15,000 linear metres of our anchors were installed from foundation to roof level with varying sizes of diamond-drilled holes to stitch the structures together.

Mostly these consisted of 20-mm stainless steel anchor bodies, installed in a 65-mm dry-diamond drilled hole. Smaller consolidation anchors, comprising 10-mm circular hollow sections in 32-mm dry-diamond drilled holes, were used mainly to link and provide horizontal connections between the inner and outer faces of the random stone walls. The white grout used was specially formulated to match the existing parent material.

To meet the requirements of the individual repairs, the anchors would vary in length from 200 mm to 20 m. For transport and logistical reasons, they had to be made on site, as delivery from the factory in the United Kingdom would have taken three months. As each diamond-drilled hole was drilled, it was precisely measured and a bespoke anchor was fabricated to fit.

The soil liquefaction, which was responsible for most of the damage done to the buildings in the earthquake was particularly evident in the historic old quarter. This was due to the broken and ineffective main drainage system which had been designed and installed early in the twentieth century to serve a population at that time of about a million. Nearly a hundred years on, that figure was approaching seventeen million, and yet the drainage system had not been updated or repaired from when it was first installed.

Thus, on Friday, the holy day in Cairo that can run into Saturday, most occupants are at home, consuming, disposing and discharging a great deal of water that inevitably finds its way to faults in the drainage system, saturating the buildings' foundations, with the consequential absorption of moisture into building fabric, and eventually rising into the external walls. The cycle is completed on Monday, when the population returns to work. The walls dry out in the high temperatures for the next few days, but not the subsoil, which remains damp.

The main contractor dictated the progress of the work. At one time we had fifteen technicians employed, working simultaneously on three major sites. The logistics and cooperation between contractors and client were always interesting, particularly when the scaffolding and working platforms were consistent with local practices, which were a million miles from United Kingdom standards.

Our personnel soon became integrated with the local population and many firm friendships were formed, including family invitations to celebrate at the many Egyptian and Muslim festivals. This included eating on a daily basis from the local fast food outlets, usually a handcart. Each technician had a local 'mate' of varying skills to assist in the work. These would also be of assistance as translators and guides to the local scene and customs.

Most of the works were completed in three and a half years, though some were extended beyond this target date as a result of revisions of the contracts.

A curious incident occurred when one of the long anchors was only half-drilled in a lower part of a large external wall in the very heart of this part of the old city. It was at the end of the day shift. The technician had removed the drilling barrel from the wall, together with all the drilling motor and frame, to store it at our lock-up. The partially drilled hole was, as usual, left open overnight. In the morning, when the

ABOVE: *Typical damage to the interior of a maqaad.*
LEFT: *Installing an anchor at roof level.*
OPPOSITE: *Minaret of the Al-Ghuri Mosque.*

RIGHT: *Interior of As-Silahdar after the strengthening work had been completed.*
BELOW: *As-Saifi after the strengthening work had been completed.*

equipment was repositioned on the wall and the drilling barrel introduced into the pre-drilled hole, a squelching noise was heard from inside it as the drill started to rotate. A very startled technician removed the drill to find the body of a large king cobra attached to the end of the drill. Obviously, the snake in this densely-populated quarter of Cairo thought it was worth the risk to look for a new home. Regrettably, it was not to be. This happened on two occasions during the contract period, and one wonders what else was lurking in all these old buildings.

ABOVE: *The approach near Bestaq Maqaad after completion of the strengthening work.*
RIGHT: As-Sarghitmish *Mosque after the strengthening work had been completed.*

3 SAVING THE TEMPLE OF HIBIS

W E WERE ALMOST THROUGH the contract to complete the work on the historic Cairo scheme when my agent showed me a newspaper cutting. In July 1999 *Al Ahram*, a weekly English-language Cairo newspaper, had reported that plans to save the Temple of Hibis were now moving ahead after fifty years of consideration. He wanted to know if I was interested in tendering for the structural project to stabilise a temple in the Western Desert. Naturally, I was delighted to be involved.

Elevation of the Temple of Hibis in the Kharga Oasis of the Western Desert.

Sketches of the temple based on the original publication by Herbert Winlock of the Metropolitan Museum, New York.

TEMPLE OF HIBIS ORIGINAL DRAWINGS

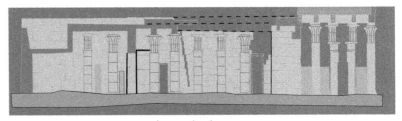

longitudinal section

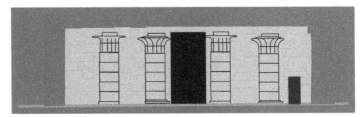

cross section

front elevation

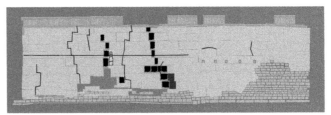

south elevation

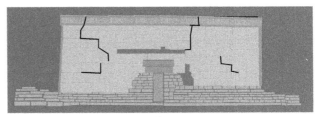

west elevation

The Temple of Hibis (the name means 'plough') is located some 700 km south of Cairo. The temple was unique at the time of construction in its abundance of inscriptions and unsurpassed depictions of the deities of ancient Egypt.

The Supreme Council of Antiquities had signed a contract with Arab Contractors (the largest civil engineering contractor in Egypt) to dismantle, reconstruct and conserve the temple, and the contract was to last thirty months. It was to be the second biggest salvage operation to have been carried out in Egypt, rivalling the work to relocate the Abu Simbel temples in the 1960s.

The Temple of Hibis had been off-limits to the public since 1980, following deterioration noticed in 1958, when it was reported that the ground water level in the oasis had risen and was endangering the structure. In 1980 attempts were made to control the rise by digging drainage channels and re-excavating the sacred lake, but this proved ineffective, and resulted in outward rotation of the external walls, movement of columns and water penetration into the masonry, damaging paintings and wall reliefs.

Supporting timbers to the temple: front entrance (FAR LEFT); *side entrance* (LEFT); *entrance* (BELOW LEFT); *and* interior (BELOW RIGHT).

FAR LEFT: *Supporting timber to the front entrance of the temple.* LEFT: *Side elevation showing raking shoring and parapet stonework.*

ABOVE: *The front entrance of the temple.*

The Supreme Council of Antiquities accordingly decided to have the temple moved to a site carefully chosen to be near an ancient quarry, so that damaged stones could easily be replaced if needed. When our agent initially asked if Arab Contractors needed our specialist services, he was told they would not be required as traditional methods for repairing the superstructure would be used.

The first task was to mark and record all the positions of the stones prior to their removal from site. This worked well, and the surveys indicated that there were no recorded problems. However, the first attempts to remove the stonework on a small ancillary structure were disastrous. The stones disintegrated as soon as they were moved, breaking into small friable elements and dust. The work was stopped in 2002 on the recommendation of the German team of Rainer Stadelmann and Hourig Sourazian.

Professor M. K. Ghamrawy, the project's senior civil engineering consultant, was then given the task of resolving the problem of the temple, and in late 2002 contacted our agents to arrange an urgent meeting to suggest a workable solution to what had become a very pressing problem. He was then directed by the Supreme Council of Antiquities to find the latest technology in civil and structural engineering to save the monument at its existing location.

Our work in restoring the historic Islamic monuments in Cairo had been well received by the client. The work to date was on time and the progress considered a success. This, together with our international record of work on landmark sites and projects, opened the door to us to provide a workable solution to the problems. It was arranged that a site visit would be made to view at first hand the damaged structure and the problem of its differential movement, with parts of the foundation rising and other parts sinking.

There was a problem of timing in getting all the design team and the general contractor to this remote location 270 km west of Luxor. The Hibis temple is the most important monument in the Western (Sahara) Desert. El-Kharga is the capital of the New Valley Governorate, which covers an area of 458,000 square miles, and is the largest but least populated governorate in Egypt. At one time there was an air service to the region, but this had stopped and had not been replaced. The only alternative was to hire a dedicated bus to transport the entire party.

RIGHT: *The front and side entrance of the temple.*
FAR RIGHT: *The unsuccessful attempt to take down the original blockwork on the front entrance porch.*

Whilst it is not a large structure, the temple is a gem to behold. The damage caused by the foundation problem was extensive, but if the subterraneous water could be controlled, we believed that, with some carefully crafted reinforcing, we could strengthen the structure without leaving any visible signs of intervention.

EUGENE CRUZ-URIBE PHD

Before we could give a comprehensive design solution to the problems, it was very important to take proper account of the temple's paintings and hieroglyphic inscriptions, both internal and on the outer walls. It was then that I was pointed in the direction of Professor Eugene Cruz-Uribe of the University of Chicago, who had studied the temple and written a detailed book on its history and archaeology. I contacted Eugene, and he agreed to give a précis of the temple's history and archaeological importance.

Europeans have known the temple since the early 1800s when a series of explorers started to compete in discovering ancient Egyptian artefacts. The French scientist and Egyptologist Frédéric Caillaud reached the Hibis temple late in 1818 and carved his name on the north reveal of the gate to Hypostyle N to proclaim his success in being the first European to go there. His main competitor, Bernardino Drovetti, an Italian agent, arrived several months later at the beginning of 1819. Over the next century, a series of European and American tourists visited the site of the temple, leaving their names at a variety of spots, but mainly on the area inside the Mani gate in the shade of the south wall. In the 1850s a Frenchman by the name of Ayme contracted with the government to build an alum factory in the temple. He arrived and began to use the temple as a quarry. In the process, he removed most of the wall (except for a small section on the south side which was under a sand dune) as well as most of the rear/west wall of the temple and the roof blocks. Only a small number of these have been recovered.

The modern excavation of the temple by Herbert Winlock of the Metropolitan Museum of Art Egyptian expedition began in 1908. Winlock brought out a large team and hired many locals to dig the site. He spent three years in the area from 1908 to 1911 doing excavations, before abandoning Kharga and transferring all of the Metropolitan's workforce into the valley at the site of Deir el-Bahri in and around the temple of Queen Hatshepsut. It appears that the excavation of the Hibis temple did not produce a sufficient number of artefacts, and the move to Deir el-Bahri yielded many items, which now adorn the Metropolitan Museum Egyptian galleries. From the Hibis temple they took only the capital from the portico. This piece now adorns the Dendur wing of the Metropolitan Museum, which also received a number of small objects from the clearance around the temple and to the north in the traditional cemeteries between Hibis and the late Christian cemetery at Bagawat.

From 1911 to 1914 Emile Baraize, a French engineer with the antiquities service, began work on the reconstruction of the temple. The task was completed before the outbreak of the First World War and proved to be an excellent accomplishment, completely restoring the fallen great gateway and portico, stabilising all the walls, rebuilding the exterior west wall and constructing a modern roof to protect the interior decorations. Baraize was also able to reconstruct Room H on the roof from a group of blocks found in the mosque, which had been torn down in the village.

In the 1920s the Metropolitan Museum team made yearly trips to the oasis to take photographs and make drawings of the decorated scenes.

The initial publication about the temple took place in 1939, when H. Evelyn White and J. Oliver published the second volume of the excavation report. This contained many of the Greek inscriptions found at the site, including several famous texts from, among others, Tiberius Julius Alexander, who was governor of Egypt in AD 69. The first volume by Winlock appeared in 1941. It was a description of the excavations in and around the temple and provided an overview, a short history of the decorations of the area and the subsidiary buildings, as well as a list of modern European graffiti and an appendix listing all the ancient Greek and Roman coins found at the site. It also provided the first comprehensive scale drawings of the temple and a small number of photographs by Harry Burton (who did the photography for the Tutankhamen tomb excavation).

The third volume was entrusted to Norman de Garis Davies, the artist and draughtsman who was in charge of publishing the decorations of numerous tombs and temples from the Theban area. He went out to Hibis on several occasions in the 1920s and 1930s, making preliminary sketches and drawings. He unfortunately died before he could complete the task, and his volume was given over to two staff members at the Metropolitan Museum who had never visited the site. They pieced together his sketches, and a large folio volume appeared in 1953. This gave sketches of most of the decorated scenes in the temple and a short description of each.

In 1984, Professor Eugene Cruz-Uribe began a field project to finish the work begun by Davies and translate all the ancient inscriptions at the temple. Funded by a grant from the National Endowment for the Humanities, he conducted three field seasons in the winter of 1984–5, the winter of 1985–6 and summer of 1986. The task of his expedition was to make corrections to Davies's drawings (only a few needed), to trace and make drawings of any scenes Davies had missed (twenty-eight scenes and fragments of scenes were copied), and to translate all the hieroglyphic inscriptions. Cruz-Uribe's first volume appeared in 1988. In it he gave translations of all the texts, drawings of the new scenes and discussions of several important topics such as the decoration, the sanctuary and Room E on the roof, which were reconstructed. His work on the temple complex continues. In 1995, he published a volume on the Demotic graffiti from the Gebel Teir quarry related to the Hibis material, and finished a volume publishing the 350 mostly ancient graffiti (text and figures) found on the temple wall.

CONSTRUCTION OF THE TEMPLE

Amasis II started the construction of the present temple in the 26th Dynasty (672–525 BC), and it was later completed by the Persian King Darius I. It is believed that this was not the only temple built on this site, which was possibly a holy one dating back to the Old and the Middle Kingdom.

During the reign of Hekr, Nekhtnebu I and Nekhtnebu II in 360 BC additions were made to the temple, and in 284 BC the external portal was constructed by Ptolemy II. The entrance to the temple is from a Roman gate created after the death of the emperor Nero in AD 68.

The temple is a long rectangular-section masonry structure approximately 44 m long and 19 m wide. The external wall is approximately 7.5 m above ground

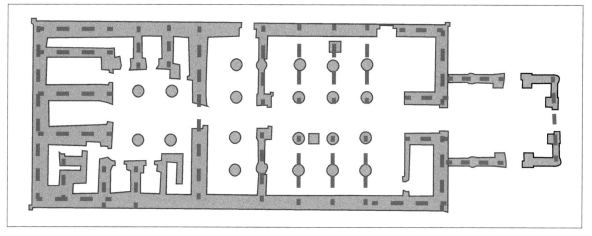

level and the foundation footings extend 2 m below. In 1910 the Metropolitan Museum of New York excavated the temple from the covering sands. The work was continued in the 1940s and 1950s, when concrete was widely used to make repairs. Two stone-faced columns were added to the portico, precast concrete slabs were placed over much of the temple area and trough sections of the same material were used as small beams or parts of the entablature.

Plan of Temple of Hibis showing anchor positions as dashed red lines.

THE SANCTUARY

The temple is covered with paintings and hieroglyphs on all walls, and the sanctuary is perhaps the most fascinating room. Originally it was composed of a false door niche. This was later built over and replaced with a register format matching the nine registers on the north and south walls.

The contents of the free walls are astonishing. Each register depicts a series of deities, totalling 359, organised in groups apparently representing major cult centres. Each register begins with a figure of the king offering gifts of water, food, wine, oil, clothing, land etc. Most of the deities on the south wall represent the major cult centres in Upper Egypt. They start with the gods from the Philae region, followed by those from Elephantine, and then progress northwards, going on with deities of Middle Egypt. In the first register, we see the various forms of Khnum of Elephantine, and on the third the seven Hathors, who were normally associated with determining a person's fate at birth. In the fifth register are representations of the deities from Hermopolis led by a child god shown emerging from the lotus flower at the moment of creation.

The west wall is composed entirely of paintings of the deities from the Theban region with myriad images of the god Amun-Re. The north wall continues a geographical list that follows Egypt's main cults, focusing on those from northern Upper Egypt and the Memphite area. On the third register is a series of figures of the progress of Astarte, including an image of the goddess riding a horse side-saddle; and in register seven we see the figure of Isis suckling a crocodile. The inscriptions at the front vary with offerings to Amun-Re as king of the gods. The deities from Lower Egypt seem to have been added in on the reveals of the door when the sanctuary was reconstructed later in the Persian period.

ABOVE: *Depiction of the 'sema-tawy' ritual.*

THE KINGSHIP ROOM

The Kingship Room is decorated with the main themes found in other decorations. On the south wall, we have numerous scenes of the king making offerings to gods, but also several with the king being led by the gods as part of the Coronation, as well as scenes of the gods Khnum and Ptah fashioning the king on a potter's wheel. This scene is unique from ancient Egypt.

The west wall is covered entirely by the image of the king on his throne with Thoth and Horus tying the cords around the base of the scene in the 'sema-tawy' ritual. In this ritual, the sign represents the trachea and lungs of the king with papyrus and lotus plants tied around it, symbolizing the people of Egypt catching their breath of life through the lungs. The ritual also signifies the unification of Upper and Lower Egypt in the person of the king. The entire north wall is covered by a long hieroglyphic text which is the acclamation of the king as Horus and identifying him with a series of gods: Shu, Geb, Osiris, Horus, Ptah-Sokar, Harsaphes, Min-Re, Min-Horus, Min, Isis, Horuskhendtyorty, Thoth, Maat, Khnum, Anubis, Anuris, Nephthys and Neith.

Located adjacent to the Sanctuary Room, the Kingship Room provided the first clues to the eventual demise of the temple caused by the ground water rise and the onset of the problem of differential ground movement. Inside the room can be seen the builders' first attempt to stabilise the roof with a column near the door opening on the south-facing wall. This was an obvious attempt to stabilise and support the roof, protecting it from collapse through weakening of the foundations.

OSIRIS COMPLEX K-K1-K2

The small chamber in the north-west corner next to the Kingship Room is a series of small rooms with a stairwell leading up to another chamber with a small pit. This room represents the rejuvenation the god Osiris in the feast of Choiak. The stairwell follows this to walls covered by two separate inscriptions. The inscriptions on the north wall give a section of chapter 146 of the Book of the Dead converted to temple ritual use. Here Osiris passes through the netherworld. The south wall inscription follows the transformation of Osiris into a sun god. This leads to the appearance of Osiris-Re in the back of the chamber above the pit, and the associated scenes on the north and south walls show Osiris worshipping the resurrected sun god in the form of Osiris-Re.

STOREROOM: ROOM I

Room I is the chamber immediately to the right as you enter Hypostyle B. It juts out from the corner and parallels the stairwell on the south side of the area. It is the only room in this part of the temple which is not inscribed on the inside. However, the series of inscriptions around the door tells us that this room is a leading storage room. It stored the linen cloth used in the various daily rituals, including the call. The room has figures of the goddess Tait shown as a snake-headed woman. The inscriptions warn all who enter the room to be ritually purified, suggesting that to touch the divine linen was reserved only for purified priests who would perform the daily dressing ritual in the sanctuary. There is a parallel to this room found with similar inscriptions at Edfu temple.

OPPOSITE: *Wall carvings at the Temple of Hibis.*

HYPOSTYLE

The four columns dominating the hypostyle provided access to the sanctuary and all the connected rooms, which often separate side chambers from direct access to the sanctuary area. Each of the four columns had been decorated with painted scenes of the king offering to the gods. While most of these are badly eroded, it is possible to trace the outlines of most of the figures on the four columns. The first figure is up on the east wall where we have a scene of the king emerging from his palace. Here we observe the name Horus the King, which is identified as Psammetichus II (595–589 BC). The other area which once had cartouches is on the reveals of the doorways. Here the name of Darius I (522–486 BC) is carved in the cartouches. When we look at the blank cartouches on the north and south walls we notice that some of them have paint in them (blue). Note especially the cartouche on the first (top) register on the south wall, where we have the king offering to Anuris and Tefnut. There the cartouche is fully painted in with the name of Darius. In a separate publication Professor Cruz-Uribe examined all of these cartouches and was able to determine that Psammetichus II was the king who built the main part of the temple, and that Darius I commissioned the reworking of the front of the temple (later covered over by Hypostyle B) along with the decoration of the jambs and reveals of the doors to the side chambers.

One architectural anomaly in Hypostyle B can be observed on the north wall. It appears that there is some damage along the course of the stone. Close inspection reveals that when the stones were put in place during construction, several, had been improperly quarried too thick. Thus, when the stones were put in position in the wall, they were levelled and adjusted using markings on the exterior side (this is clear when you go to the roof and examine these blocks). After the work was done and the decoration of the interior walls began, it was discovered that these blocks projected 3 cm above the lower course of stones. The workers then started to try and chisel the stone but quickly stopped after realizing how much work the job would entail. They then simply chiselled off the lower section roughly and covered the area with white plaster, in which they added the carvings and other decorations. Some of this plaster still exists on the wall.

ANCIENT RECONSTRUCTION OF HYPOSTYLE B

The next topic to be discussed is the issue of the ancient repairs to the temple. In Room L there is a column just inside the door, which blocks part of the decoration on the south wall. This column was part of the major reconstruction effort that was made, probably during the Ptolemaic period (conjecture based on the style of the capitals in Room L). For some reason (either an earthquake or dramatic settling of the western end of the wall of the temple), a large group of cracks appeared through Hypostyle B. These can be seen on the inside and outside of the temple running along the north–south line through the doors of Rooms K and F.

Around the door to Room K on the north side of Hypostyle B we can see a series of blocks that have been put there to replace the damage to the reveal/jamb of the door. There are no carved decorations, but only traces of paint on the surface. On the door to Room F (south side of Hypostyle B) we can see major reconstruction. Portions of the door jamb/reveal have been chiselled out and would have been

OPPOSITE: *Palmiform columns at the Temple of Hibis.*

filled with plaster. In addition, a blank block without decoration has replaced the lintel of the door. Above both of these doors one can see the repairs made to the other blocks which were damaged in antiquity.

By examining the exterior walls at various places along the north–south axis, you will also see that repairs were made in antiquity. A series of blocks, all decorated, have been fitted into the exterior wall at these positions, indicating a repair of damaged or destroyed stones. This was first noted in the elevation drawing provided as plate 35 in the Metropolitan's archaeological report, with this section shaded in grey.

There are several other areas on the north and south exterior walls that show the same ancient repairs. It seems quite clear that at some point, probably during the Ptolemaic period, the temple suffered severe damage and was repaired by local craftsmen. Their work mainly consisted of replacing damaged blocks both on the interior and exterior with blank and decorated blocks. In those areas that had decoration they replaced the painting but did not carve any of the hieroglyphics. Some of this painted decoration from these repairs can still be seen, especially on the door jamb on the door to Room K.

THE RESTORATION

The main underlying problem with the temple was the water incursion into the foundations causing some sections to settle and others to lift. The significant rotation in the foundations caused by the water was accelerated by adjacent cultivation and irrigation. There was a great deal of fracturing of the sandstone blocks, creating voids in many of the walls. Some had become unstable as a consequence. In previous repair phases, several sandstone blocks had been replaced with precast stone-faced concrete elements. Some of the surviving entablature stones had fractured and these had temporarily been secured in position with steelwork. They were in need of repair. The external walls had suffered most through the change in moisture content of the soils on the underside of the footings. The walls had started to rotate outwards and temporary timber shorings had been introduced to

The distressed condition of the temple prior to renovation work.

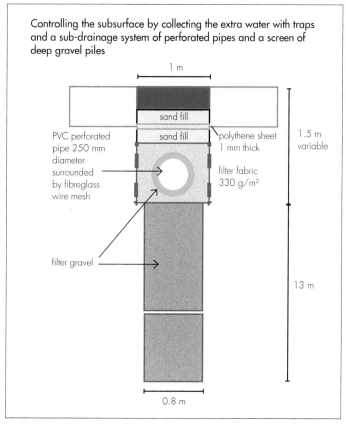

Controlling the subsurface by collecting the extra water with traps and a sub-drainage system of perforated pipes and a screen of deep gravel piles

1 m

sand fill

PVC perforated
pipe 250 mm
diameter
surrounded
by fibreglass
wire mesh

sand fill

polythene sheet
1 mm thick

1.5 m
variable

filter fabric
330 g/m²

filter gravel

13 m

0.8 m

prevent further movement. Professor Ghamrawy had devised a scheme to isolate the structure from further moisture incursion. His solution was to excavate a metre-wide trench 16 m deep around the entire temple complex. The trench incorporated a new drainage system and a fabric filter on the bed of the gravel. Within the temple complex a series of lime-filled piles was installed, again at a depth of 16 m, at metre intervals. This was to control and harness the swelling potential of the foundation soils and absorb any surplus water.

Our technical staff together with Richard Swift, our consulting engineer, addressed the problem of the damage to the superstructure. In order to stabilise the external walls, it was deemed appropriate to use a system which disturbs the existing structure as little as possible, yet achieves stability by connecting the external walls to the internal arrangement of the structure. Richard Swift and SFK consulting engineers have long experience working with Cintec on similar structural problems and knew of the advantages of using a product that is generally invisible within the structure of the building and imposes very low levels of stress on weak masonry. It may be designed so that the size of the anchor body reflects the strength of the masonry. It can be installed in long lengths, and the materials used in the anchors are not deleterious. As it was presumed that the foundations of the external walls were rotating about their inner edge, lateral restraint anchors to the external wall connecting to the internal walls were considered desirable.

TOP LEFT: *Special excavating equipment cutting the 16-m deep barrier to receive the stone filling.*

BOTTOM LEFT: *The 16-m deep trench was constructed around the entire perimeter of the temple and filled with aggregate.*

TOP RIGHT: *Section through the 14.5-m deep drainage barrier used to prevent water penetrating the temple foundations.*

The higher the anchor is positioned above the foundations the greater the effectiveness of the system, as only small force is needed to prevent the rotation. But additionally, the walls need to be made to act as a single element rather than separate blocks. Introduction of short consolidation anchors between the inner and outer skins of masonry similar to wall ties, as well as the more substantial anchors connecting internal and external walls, achieves the best results in consolidating external and masonry wall structure.

The surviving remains of the temple are in remarkably good condition considering its immediate past history, although, a little fractured and with some of the structural elements rotating slightly.

ENGINEERING CONSERVATION PRINCIPLES

The aim of the conservation programme was to preserve as much of the existing fabric of the building as possible. This posed several questions for further discussion with the Supreme Council of Antiquities: Were the main elements of the structure fit for purpose? If so, it may be considered more acceptable to leave these elements in place, even the recent concrete roof section, because its removal would cause more damage to the adjacent historic elements. The removal of the unsightly temporary timber raking shores was regarded as essential, both to limit the intervention as much as possible and to ensure the continued survival of the temple. Only the absolutely essential structural work to conserve the monument was to be undertaken, because future generations will undoubtedly have better methods of conservation and we should not seek to repair every part of the temple.

The presence of hieroglyphics on the external wall face of the temple restricted the possible entry points for anchor installation. One means of avoiding drilling to the face of the external wall, yet still creating an effective connection for the individual stones of the external wall, was to drill the anchors from the top of the wall at a downwards angle connecting all the damaged stonework. These anchors, together with some localised short consolidation anchors, would increase the stiffness of the walls. The lateral element of tying is then provided by the connection of anchors installed in the tablature, above the columns.

The corners of the buildings were held by inclined anchors to the vertical plane, relating to similar anchors on the inner walls. Calculations were made of the overturning moments of the walls and the necessary restraints to prevent this. Fortunately, there were many intersecting internal walls, which could be used to restrain further outward movements. The existing concrete roof structure could be considered to act as a horizontal plate diaphragm.

The location of the roof slab on top of the external walls meant that anchors installed through the external wall into the roof slab would be the most efficient way to prevent further outward movement. Many of these walls were constructed of two skins of masonry which were splitting apart.

It was important for stability purposes that the external wall should be made to act as one unit. Our consolidation anchors were of small diameter, and when introduced between inner and outer skins they provided stability.

It was considered that all the diamond drilling should be dry to prevent moisture from penetrating into the decorated faces. No attempt was made to tie

Drilling to avoid hieroglyphs, and installing anchors in temperatures of over 40°C. Ice was obtained from a local hotel to cool the water when mixing the grout.

Carefully checking the socked anchor before placing in the drilled hole.

An internal wall anchored at high level.

Carefully positioning and diamond drilling core holes in the columns to avoid the hieroglyphs.

ABOVE, TOP AND BOTTOM: *Injecting grout into the body of the anchor.*
LEFT: *Wall decorations located inside the temple.*

across the widening fissures, which had occurred in the length of the wall. These created natural movement joints, and if stitching had been attempted, new cracks would have appeared elsewhere. These cracks would be simply filled with a suitable mortar – as will be frequently required in future times. Fractured large stones were repaired, again with our stitching anchor, and this method was used on lintels on the south side of the temple columns in the north area.

A subcontract was formulated in late 2005 between our agents in Cairo, Intro Trading and Arab Contractors for the first part of the two-stage contract to install Cintec anchors in the external walls of the temple. Richard Swift had prepared detailed drawings from the original drawings supplied by Arab Contractors and the Supreme Council of Antiquities.

Making good the drilled hole to match the original substrate.

Owing to the amount of movement in the walls since the original drawings were made, it was not possible to record the exact position of the anchors on the working drawings. Measuring the movement of the blocks, their locations and the varied angles to install the anchors was undertaken on a daily basis by our specialist team as the work proceeded, and was accurately recorded. Frequent telephone calls were made between Richard and Dennis Lee, our project manager, regarding the exact location of the anchors, so that the positions could be varied slightly to avoid hieroglyphs and overall faults in the external walls. Dennis was then able to direct the drilling operation to meet the correct intervention in keeping with the overall design requirements. At the end of each week he would liaise with the Arab Contractors site engineer, and between them they would mark up on a laptop computer the exact location, position and length of each of the major anchors.

The first part of the drilling operation required great skill and care to avoid damaging the structure, surface paintings and hieroglyphs. This was particularly difficult because of the dust created by the diamond drilling undertaken in temperatures up to 45°C. The main problem during the installation of the anchors was to keep the grout mixing water within the design temperatures. It was necessary to have a daily taxi run between the local hotel and the site with large quantities of ice. The other hazard was security, and the authorities provided an armed guard present on site and at the hotel at night throughout the contract.

4 THE RED PYRAMID

I N JUNE 2003, together with engineer Emad from Intro Trading Cairo, I had a meeting with Dr D. Shereif, a consultant engineer working for the Supreme Council of Antiquities, to discuss a small problem with the interconnecting corridor between the two ante chambers of the Red Pyramid at Dahshur. I was given drawings of the chambers and details of the large cracks in the granite stones above the interconnecting corridor of the entrance to the burial chambers.

Dr Shereif requested a Cintec solution to the problem as soon as possible, so that he could present this to the Supreme Council for approval. I was more than happy to oblige; in fact, I was delighted. It had always been my ambition to install one of our anchors into a pyramid, and this opportunity was a dream come true. After various meetings with Richard Swift I was able to provide a solution to the problem within a month for Dr Shereif to consider and pass on to the authorities.

At that time, my knowledge of pyramids was very sparse and I had to consult the numerous publications by many distinguished archaeologists on the construction and chronology. It was then that I discovered the name of the Pharaoh Snefru and his family who, it is claimed, built 70 per cent of all the large pyramids near Cairo.

Pharaoh Snefru acceded the throne in 2575 BC and was one of Manetho's Fourth Dynasty of the Old Kingdom. The Step Pyramid of Djoser was then the only large royal pyramid standing, and it is believed that he must have been present during its construction. Excluding the disputed pyramid at Meidum, Snefru's second pyramid is located in Dahshur, some four miles north of this first pyramid, and is known as the Bent or North Pyramid.

The Red Pyramid derives its name from the colour of the core stones, which are a brown shade of red. The local inhabitants know it as the Bat Pyramid. An Englishman named Edward Melton is recorded as having visited it in 1660. The first real archaeological investigation was carried out by Rainer Stadelmann in the early 1980s. This was the first true pyramid ever built. Measuring 220 m x 220 m, it is 105 m high with a base angle of 43°. What is significant is that this mirrors the revised shallow angle of the top half of the Bent Pyramid, so that it is generally accepted that the Red Pyramid was built later.

It would appear that Snefru was unwilling to have his mortal remains interred in the burial chamber of the Bent Pyramid because its walls were closing up under pressure from the poorly designed corbel ceiling. Corbels are used to project as offsets or cantilevers beyond the face of a wall. There are set structural rules governing the amount of cantilever and the load they can carry. In this instance, the corbelling was high up on both sides of the burial chamber, forming a roughly vaulted arched ceiling. If the cantilever offset distance on each of the corbels has been miscalculated the walls of the burial chamber slide together under the enormous pressure, closing up the internal dimensions of the burial chamber. This was also the reason why heavy cedar beams, almost certainly added to brace the walls and keep them apart, were discovered in the Bent Pyramid burial chamber, and why the builders lowered the external construction angle from 54° to 43° halfway through the project, hoping that, by reducing the weight imposed on the burial chamber, they would reduce the pressure closing the walls together.

The visible core blocks of red limestone on the Red Pyramid came from a stone quarry some hundred metres away. It was observed that two short ramps lead from the quarry to the south-west side of the pyramid.

The entrance to the pyramid is located in the north wall, and is about 28 m above the ground. The entrance to the burial chambers is down a steeply descending corridor approximately 70 m long that coincides with the ground floor level of the pyramid. The first so-called antechamber is a short passageway that connects to a second antechamber aligned with the pyramid's vertical axis. Both chambers have the same dimensions and are built of granite, with corbel-vaulted ceilings of dressed stone.

Elevation of the Red or North Pyramid.

It is significant, that the builders took heed of the problems on the Bent Pyramid and the cantilever on the corbel is greatly reduced, giving a symmetrical appearance to the corbel construction.

The problem we were faced with was not on the exterior of the pyramid but on the corridor linking the two antechambers. The granite beams spanning the entrance via a low corridor were badly cracked from the base of the stone beam, up through the centre to a position just adjacent to the centre of the top.

Whilst the work was in planning, we had a tentative enquiry from a company

Entrance to the Red Pyramid.

in Cardiff called Green Bay Film. They had had a hint that *National Geographic* was interested in pyramid projects in Cairo and wanted to know if I would participate in a taster documentary film showing our work to repair the damaged stonework, to be later shown at the Cannes Film Festival to see if it was marketable.

This was agreed and arrangements were made to tie in the works programme with the filming schedule. Arrangements were made to have the project started in May 2006 when the setting out and drilling were to begin.

The project was not large by any standards, but the building was a pyramid and the work was to be showcased, so it had to be faultless. The crack could not have been in a more awkward position. It spanned the corridor between the two burial antechambers so that it could not be drilled in the usual way. It was impossible to drill at right angles to the corridor because of the mass of core in the pyramid. It was, however, possible to drill at an angle of 43° in each of the chambers at the entrance and exit of the corridor. We were thus able to reinforce and secure the beam with a row of 20-mm diameter stainless steel consolidation anchors diamond drilled in a 40-mm hole.

PROBLEMS OF THE OPERATION

Because of the low-lying position of the burial chamber in the monument and its distance from the surface, it had to be closed to the public. This was also essential because of the amount of dust generated using dry-drilling equipment and the impossibility of extracting it from the atmosphere at such a distance. All technicians had to wear breathing masks whilst drilling the holes.

Then a major problem was encountered in drilling the interior granite walls. These were lined with granite and the cutting segments on the diamond drill were not making any impression on the hard surface. After four hours of continual hand drilling, creating copious amounts of dust, we had managed to penetrate only 50 mm into the granite surface.

Hearing the news in our Newport head office, I changed my plan to travel to Cairo a few days later, after the drilling operation had gained momentum. Explaining the problem to the drill tip manufacturer, I requested a different set of drill tips. The

general rule is: the harder the material you are drilling, the softer the matrix used to hold the diamonds in place, and vice versa. The new core drills were couriered overnight to Newport, and I was able to advance my travel arrangements and fly to Cairo with the new core bits in my luggage. The cutting speed was better on the new core bits, but it was nowhere near what we normally expect. So, in agreement with the authorities, we were able to introduce a small amount of wet air into the process. That finally did the trick.

Curiously, we had more problems associated with the project than normal, and this was quite apart from the work itself. The film director wanted some atmospheric shots of Cairo to provide a more colourful backdrop to the film. The cameraman and the director were in the rear seats of a car travelling through the centre of Cairo. I was in the front passenger seat, wearing a new shirt freshly put on that morning. But the director wanted me to be in the same shirt I had worn on the day before for the sake of continuity. As luck would have it, my case was in the boot of the car with all my dirty washing, but I was reluctant to change my shirt in the middle of Cairo, in view of the

Inside the Red Pyramid.

strong moral attitude observed in Egypt. Unbelievably, I was persuaded to put the old shirt on top of the one I was wearing. As we were travelling towards the Cairo Museum in Tahir Square, the car was forced to the nearside kerb by a medium-size panel van. As we stopped, about half a dozen armed men dressed in black burst out of the van and quickly surrounded the car. They demanded to know what we were doing, and did so in a hostile and threatening manner. When we had shown our documents and proved that we had official permission to film, they were happy with our explanations, but they wanted to know if the idiot with two shirts on in the middle of the day in boiling-heat Cairo was sane.

A little later, we visited the Mosque of Al-Ghuri, our first-ever Cairo project in 1998. Again, this was for background footage as it was where we cut our teeth in Egypt.

When we first started the project in 1998 an ever-present elderly lady was standing, it seemed permanently, outside the mosque. Out of curiosity she asked me what we were doing. When I explained that we were going to save the structure of the mosque, she said 'God bless us.' I was thankful that when we visited the mosque again to film a few frames, the same lady was outside as she had been eight years before. Obviously, she did not recognize me, because as soon as the camera was positioned outside the main building and the director and audio assistant started the filming, she started shouting that we should not be there. The cry was taken up by a number of other people and what was almost a mini-riot was developing. Luckily, a passing policeman was able to control the situation, and we moved at a great pace from the mosque, as in this instance discretion was far better than any valour. I decided that that was more than enough colour for that day.

5 HOW THE PYRAMIDS WERE BUILT, THEIR RISE AND FALL

M Y FIRST SIGHT OF THE Great Pyramid of Khufu on the Giza plateau through the trees has always stayed in my memory. The sheer scale of the monument eclipses all imagination or reasoning to explain how it was constructed. The human effort and the then new technology needed to raise this structure to a height of 146+ metres were breathtaking. The logistical support required to move the vast amounts of materials, the infrastructure, the skilled labour and the courage to reach where no one had gone before are staggering. But why a pyramid? Why did it need to be so big? Did it need to be of this scale and height for the returning spirit to find its way back to its mummified body after visiting the stars? The question is still open.

Having been involved with the repairs to the Step Pyramid since 2006, and prior to that with many projects on the structural restoration of ancient Cairo, the Temple of Hibis and the Red Pyramid, I have to put myself in the shoes of those ancient innovators who, through trial and error, evolved and constructed these magnificent monuments.

Obviously, the pharaoh was the guiding light in planning with the priests for the afterlife and the particular need to build a bridge to the stars and also a terminal that could be seen from stars back to earth to guide the pharaoh back to his mortal remains.

The first question must be who controlled the project? Was it a single royal overseer such as Imhotep, who built the Step Pyramid, or a panel of building experts who were the leading authority on large-scale construction and turned the pharaoh's instructions into reality?

From my perspective, reviewing the results and following the progression of the construction from the first to the last of the large pyramids, it appears that there must have been a moving spirit that linked the whole series of pyramid buildings. Starting at the Step Pyramid through to the final Great Pyramid, one can see the link between the design faults and the attempts to rectify the problems. I feel the highest admiration for these ancient builders for their innovation and skills in overcoming the difficulties and changing the way engineering problems were resolved.

Consulting the vast array of archaeological information on all aspects of the building of these structures, some in great detail on the alignment, setting out, quarrying, transporting, and the theory of how they were constructed, each adding to what has been written before, I will not comment on these assumptions in detail because the cause and effect will become clear when the method of construction is analysed.

Mankind's rise out of the ground must start with the Step Pyramid. It has been acknowledged and confirmed by most authorities that this is the earliest large structure above the horizon. Dating back 4,700 years, it was at the cutting edge of building and civil engineering technology in ancient times.

Unless we build a time machine capable of whisking us back all those centuries to examine the builders as they were constructing the pyramids, one can only guess at how these extraordinary structures were built by examining the evidence, both archaeological and structural, to determine methods and innovations required. We must approach the problem in the same manner as the great Sherlock Holmes, whose maxim was that if you remove all evidence that doesn't fit the facts, then what's left, must be the way it happened. What is not in doubt is that the pyramids still stand, and can be seen, and each tells us its own story. Even though, after all

The Step Pyramid of Djoser, Saqqara.

these centuries, they are no longer in their original state, their present condition provides us with an insight into their mode of failure.

We have some knowledge of the problem, however partial and unreliable, from two ancient Greek historians. Herodotus, one of the earliest recorded travellers to be captivated by Egypt, visited the country in the fifth century BC, probably in the course of collecting material for his *Histories*, which provide the first known narrative of how the pyramids were built. Herodotus relied on an oral account, allegedly from a priest, but with very little in the way of technical information. The other source from antiquity is Diodorus Siculus' *Bibliotheca Historica*, written between 60 and 30 BC.

Herodotus recorded that the Great Pyramid in Giza took 100,000 men and twenty years to construct. At first, in his account, the pyramid was built with steps, like a staircase: Once the first step had been built, the stones needed for the next were lifted by means of a short wooden scaffold. They were laid on another scaffold, by means of which they were raised to the second step, and so on. Lifting devices were provided for each step, in case the scaffolding was too heavy to be easily moved from step to step once the stones had been placed in position.

Diodorus Siculus says that the stone was brought over a great distance, and that the construction was undertaken with the help of ramps. These very different accounts provide the basis for modern approaches to the problem. Some researchers, relying on Herodotus, assume that the stone blocks were raised with the help of simple wooden structures; those who follow Diodorus maintain that massive, elaborate ramps were used.

It is also possible that the diversity of the accounts results from observers at different periods having different views or aspects of the work and during the erection of the core stones observing different methods of construction. Certainly, both scaffolding and ramps were used.

There are other theories that have no basis in science or engineering, or rely on techniques that were not available to the ancient Egyptians or any people from this planet. For my own part, after years in the navy, I come back to the service's maxim of KISS ('Keep it simple, stupid').

One thing is indisputable: it took a very long time and a great deal of effort to construct these vast monuments, periods of between ten and twenty-five years for each pyramid. Using only primitive copper tools and muscle power, and with very little else, the builders successfully completed the world's largest group of structures which were unsurpassed for many centuries. Measured by today's standards, the achievement of going to the moon is a comparable equivalent.

The initial instruction, to provide a suitable vehicle which would lift the God on Earth, the pharaoh, to the afterlife, must have come from the pharaoh himself and his priests. The Royal Builder would be consulted and would formulate a design relying on the past experience and failures in building earlier pyramids. There was probably an organization of master builders comprising what we would now know as architects, surveyors, engineers, stonemasons, carpenters and scaffolders, together with ancillary skilled workers such as scribes and logistical providers, including the boat builders and river navigators who would need to be consulted to determine that the work was feasible. In particular, the overseer would need to consult the grey-haired members of previous teams who would have had experience with the problems encountered.

TOP: *A view from the top of the* Step Pyramid. *In the distance, the Red and Bent Pyramids are just visible.*

BOTTOM: *More traditional rope and wooden scaffolding is required as the renovation work proceeds.*

The cause of failures at the Step Pyramid, the Meidum Pyramid and the Bent Pyramid would have provided the builders with knowledge of how not to build the next pyramid. All aspects of the failed designs and all the lessons learned would have been carefully considered and mistakes avoided in the next project. The accumulation of failures was fundamental to shaping the building plans for the Red Pyramid and those that followed.

The Royal Master Builder and his team would have examined the results of the failures and successes, addressing them in the new designs for the next generation of structures from the Red Pyramid on to the Giza Plateau; they would examine these in chronological order to establish a pattern of how they incrementally, addressed the issues, applying the lessons learned to the next structure to be built.

I start with the Step Pyramid. This was ground-breaking, not only because it was the first successful structure to be raised to an astounding 62 metres above the ground, at a time when no other building had reached even a quarter of this height, but because the structure was made from stone and not the traditional clay bricks used on mastabas up until this time.

In my view, the reason the pyramid is stepped is that the builders had no knowledge of how to construct scaffolding above 11 metres in height. Imagine, that you are using wooden scaffolding successfully erected to a height of 11 metres. The wooden standards and ledgers are made from roughly hewn timber and joined with rope lashings. To construct these to a height of 62 metres would have been a daunting task for builders who had never experienced these heights. To overcome this problem, which probably also related to the stonework as well as the scaffolding at those heights, the construction was stepped like a wedding cake in profile. As each layer was built, the next layer was constructed, but stepped further back from the outer perimeter of the base, creating a new small footprint on top of the already completed lower tier. This worked exceptionally well; it produced the magnificent structure that we see today, despite the ravages of weather, lack of maintenance and the constant pressure of gravity.

But, there was a latent defect in the design of the pyramid that affected its essential purpose. This not only dogged the Step Pyramid, but also the next two pyramids in succession. How do you create and span a large area inside a pyramid and support it to withstand the force of gravity? The loads borne by these colossal structures are enormous, and prior to the Great Pyramid there were no stone beams long enough to span the required openings. Another way would have to be found.

The solution must have started with the stonemasons, who discovered that if they laid each course of stones to overhang the course below by a small distance, and if this process was reflected on the facing side of the structure, it could create an air space inside the structure, just as is done by a simple arch. This is known as corbelling. The first use of corbelling was in a burial chamber in the pyramid at Meidum. This was the next in line after the Step Pyramid, and attributed either to the Pharaoh Snefru or his father Huni.

I have to wonder, why they attempted this form of opening in the pyramid, straight after the timber ceiling in the Step Pyramid. How did they know that the timber ceiling in the Step Pyramid was inadequate? When did the builders discover this? Before the Step Pyramid was completed? Or, later when it was robbed? Or before the Meidum Pyramid was started?

It is generally believed, that the tomb robbers were the builders themselves (and having been in this industry for many years I cannot disagree). There are two explanations for this anomaly. The first is that the burial chamber in the Step Pyramid was not completed but began to collapse before or at the same time that the pharaoh was buried, for all to see and observe. Alternatively, the pharaoh was buried with all due ceremony as befitted his station, and at a later time the tomb was robbed and the culprits observed the collapse in the ceiling, before the work on the next pyramid at Meidum was begun.

If so, how was the information transmitted to the Royal Master Builder that the timber ceiling in the burial chamber of the Step Pyramid had collapsed? Surely, if it were passed to a government official, it would be signing one's own death warrant to admit that one has entered a sacred chamber and desecrated a tomb?

Yet if he had not known that the Step Pyramid burial chamber had collapsed, why did the Royal Master Builder radically change the design on the pyramid at Meidum? Surely, the Meidum Pyramid was not started until well after the Step Pyramid had been completed.

To use corbelling on the Step Pyramid, the Royal Master Builder would have needed to dramatically change the design of the surrounding stonework. Corbelling has to be installed in courses as part and parcel of the burial chamber walls. The load has to be transferred into the structure, and the method cannot be used in the random unbonded stonework that we discovered surrounding the burial chamber ceiling at Saqqara.

Corbelling was first attempted at the next pyramid at Meidum, located some 50 kilometres due south of Dahshur. However, the method of construction was changed. A central core of stones was needed to make the corbelling experiment possible, the pyramid shape being added afterwards on all four sides of the central tower using eight stone-clad layers 4–5 metres wide at 76°, like vertical onion layers, a method using what is known as accretion theory.

It was also necessary to locate the burial chamber near ground level, not underground as in the Step Pyramid. This would make sense, as they would need to have 360° access to the chamber to construct a corbelled ceiling.

The first prototype, constructed of limestone blocks, was a very rough and ready attempt at corbelling. It was not symmetrical or accurately formed. Some corbels differed in size from those above and below and the cantilevered distances also varied from course to course, but this was an experiment which provided the builders with the insight that would lead eventually to success in the pyramids to follow.

Some commentators have remarked that the stone blocks were in their undressed state and not finished. Corbelling is a structural technique, and any attempt to dress or change the appearance would affect the load-bearing capacity of the system and would not be advisable.

That was probably, the only good innovation discovered on the Meidum Pyramid. If one measures failures as a guide to future success, then this structure has them in abundance. It is a complete ruin, with only the central core tower poking up through a sea of rubble at the base, giving an impression of a violent attack with a huge sledgehammer on its outer facings.

Kurt Mendelssohn reports in his excellent book *The Riddle of the Pyramids* that the first European traveller described how he saw the monument. This was Captain Nordon, who had journeyed to the site and made three excellent sketches of the pyramid from a distance. They show the building in its ruined state, very much as we see it today, except that the rubble surrounding it seems to be a little, but not very much, higher.

That was in 1737, and a little later in the same year another Fellow of the Royal Society, Edward Pococke, also made a note of seeing the pyramid. The next visitor was W. G. Browne who, in 1793, explored the actual site and, took into the debris some casing stones of the pyramid. He concluded correctly that the tower was not standing on a natural hill, but that it was the rubble surrounding the building, which gave this impression.

At the time of the Napoleonic invasion of Egypt, Denon, a French scientist, visited the pyramid, and produced drawings which he included in the expedition's records. He was followed by the renowned Egyptologists, J. S. Perring in 1837, K. R. Lepsius in 1843 and Auguste Mariette in 1871. Perring provided an excellent sketch of the building and suggested that some of the casing blocks had been used to build the bridge at Tahme.

Flinders Petrie, on his first visit to the pyramid, thought that stone robbers had caused the damage. He returned in 1891 and discovered that there were three distinct building stages. The first two of these were step pyramids, one of seven steps, the next probably of eight. Only, the second step of the pyramid was covered with an outer casing, of which only the lowest part now remains.

The next two steps, the third and fourth, have collapsed, leaving the core of the fifth step standing, surmounted by the sixth and parts of the seventh step. What is left of the outer casing indicates that the third building phase was to be a true pyramid, and this view has been generally accepted. Petrie, calls the first step pyramid E1, the second step pyramid E2, and the outer mantle of the true pyramid E3. Again, this nomenclature is generally used in discussing the construction of the monument.

In 1909 Petrie returned to the pyramid with a team of specialists which included the architect George Fraser, the Egyptologist Percy Newberry, and the

CORBELS

Brick corbels
Modern practice is usually a cantilever of no more than one-third of the mason unit.

Corbels are usually used either as a decorative architectural feature or to provide support for the overhanging detail. Their use is normally only for a small number of steps.

The corbels used to form the openings in the Meidum and Bent Pyramids were uneven and with a cantilever too great for the imposed load.

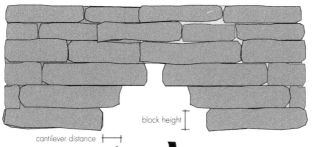

block height

cantilever distance

The opening formed with corbelling in the Bent Pyramid could not support the load imposed by the steep angle of construction. The corbel stone blocks were not restrained and slid together closing the opening.

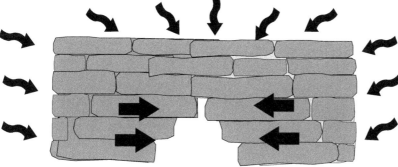

section through
Grand Gallery

The Grand Gallery chamber of the Great Pyramid leads to the burial chamber and is 47 m long by 8.5 m high. The extreme loading imposed by the size and weight of the pyramid on the opening has been compensated by the use of small cantilever corbels.

Corbelling renders the walls less stable by bringing the centre of gravity of the mass away from the centre of the wall, causing greater compression on the side of the wall under load than on the side remote from the load. The distance of the centre of the bearing is known as an eccentricity of the load. If this is too great, it leads to a sideways thrust with a tendency to rotate and disintegrate the wall.

archaeologists Ernest Mackay and Gerald Wainright. They drove a tunnel inwards towards the centre of the pyramid, underneath the base and through the outer casing (E3), finishing at the rock foundation of the burial chamber. Petrie found the foundations of ten successive buttress walls. This proved that the pyramid was not a mastaba like the Step Pyramid

They found that the outermost buttress wall of the second step pyramid, E2, beneath the outer casing was fair-faced and finished almost to give the impression that it was the outer layer. However, the builders were not finished, but probably intended to add further steps, as on the original Djoser Pyramid. The final stage before the transformation of the monument into a true pyramid (E3) was then added. Unfortunately, the construction of these unbonded layers was not conducive to the stability of the pyramid as a whole.

Further proof that the two step pyramids E1 and E2 were for a time regarded as the finished article before the next phase was started was provided by Petrie during observations in a passage descending into the pyramid. The internal masonry lining of this corridor shows clear discontinuities at those places, which correspond to the original entrance of the successive step pyramids, E1 and E2. Beyond these, a passage was subsequently continued outwards to the final entrance of pyramid layer E3.

Ludwig Borchardt carried out the next exploration, in 1926. His theory was that the exposed courses of masonry were in fact, part of the second step pyramid, E2, laid on top of the steps of E1, as the dimensions and height of the new steps did not coincide with the first step heights, exposed after the pyramid had dramatically collapsed.

The external elevation of the Meidum Pyramid.

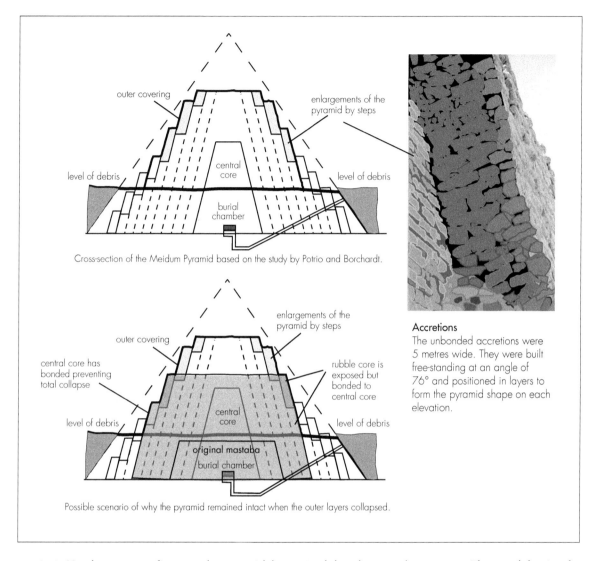

Cross-section of the Meidum Pyramid based on the study by Potrio and Borchardt.

Possible scenario of why the pyramid remained intact when the outer layers collapsed.

Accretions

The unbonded accretions were 5 metres wide. They were built free-standing at an angle of 76° and positioned in layers to form the pyramid shape on each elevation.

In 1793, when Browne first saw the pyramid, he noticed that the outer layer was completely covered with debris. He removed the rubble from two of the corners and discovered the existence of the outer casing, which he found completely intact. He also observed that the stones and the grout mortar could be seen at the base and had not been stolen by locals. There were only two small areas where the outer casing of E3 could be seen, the rest being covered by debris. None of these early examinations of the pyramid or the stone debris gave any impression that a large amount of material had been stolen from the site.

The evidence shows that whatever calamity overtook the building, it was immediate and catastrophic, with the outer layers cascading down from a great height, following an implosion near the base of the pyramid and within the layer structure.

A considerable number of mastabas constructed in the area also demonstrate the sudden abandonment of the pyramid. These were built for courtiers,

The external elevation of the Meidum Pyramid showing the collapsed accretions in smithereens around the core of the pyramid.

were never occupied and were left unfinished. It is significant that there were no tombs for mortuary priests, who usually liked to be buried close to the pyramid where they officiated.

The evidence therefore clearly points to a disaster that happened during the third building phase of the monument. Traces of mortar can be seen to the full 60-metre height of the wall, meaning that the bonding of the layers was violently ruptured by lateral forces. It is now generally believed that the inner layers were completed before the outer ones were built. The smoothly dressed surfaces of E1 and E2 give the impression that the finish of each layer was dictated by aesthetic, not practical, reasons, and thus were not bonded together, which would have been vital for the stability of the pyramid.

The tower now rises to some 62 metres above ground level and has (probably) a base width of over 130 metres square. Although, it is difficult to establish the exact dimensions of the base because of the condition of the base area, it was presumed that the central core had not been completed.

The pyramid is aligned north–south with the entrance chamber on the north face, leading to a descending passageway 57 metres long down into the burial chamber at near ground level.

The use of a form of step construction was obviously the way the builders planned the work, copying the basic format of the Step Pyramid but using new ideas and techniques that would make the monument bigger and more inspiring. It cannot be confirmed whether or not the sloping layers were provided from day one

Corbelling inside the Meidum Pyramid.

of the construction or were added incrementally after the central core or part of the central core was built. Whatever way it happened, it was to prove the downfall of the structure.

The real disaster was the outer accretion layers that clad the central core of the pyramid, changing it from a step construction to a true pyramid shape. Probably, the most revealing information on the construction of the cladding was given by Petrie and his party. Over several visits, they were able to establish, and discovered by tunnelling, that the core is inclined at an angle of 76° of semi-coursed stone composed of five accretion layers.

However, such a small sample of stonework at the base of the pyramid may not represent how the other layers were constructed. Indeed, the amount of fine debris strewn at the base of the pyramid, would lead me to think that perhaps, these layers were not built completely of coursed stone but could have been built with semi-coursed random stone. The ancient mason would have used selected stone on the front facing of the layer and infilled with smaller unselected stones at the rear. These would all be roughly bonded to the core or, if it was an additional layer, to the preceding internal layer. This was probably to speed up the construction and in any event little mortar was used to bond the matrix together and provide a homogeneous mass.

The evidence that the outer layer E3 is built on a foundation of sand is also a damning indictment of the catastrophic problems to follow. In a modern building, the outer cladding is secured to the main structure using materials such as steel ties that are able to transfer the load from the panel into the structure. As this solution was not available to the ancient Egyptians, they did what they could to key in the outer layers using masonry courses between the steps.

From the outside of the remaining tower there are some smooth sections of wall and there are others that show a band of stones projecting from the face of the tower. This was obviously where the inner core of the tower was bonded to the outer accretion layers. This connection using stone and mortar would not be sufficient to transfer the weight of the layers into the main body of the tower. Stones and bricks in compression (when a direct load is on top pressing down) are very brittle and will shear easily if unsupported, particularly, if the load is eccentric (not acting directly above but at an angle), as it would have been at 76° in this case. Even a horizontal tie connection of stone and mortar will have very little sheer value in these circumstances.

As the outer accretion layers are built, and rise up the side of the central core of the pyramid, their only means of support is the friction created by resting on the 76° angles on the side of the pyramid's main tower, or against an inner accretion layer. These outer accretion layers, in the view of Ludwig Borchardt, confirmed that they were added to the central core on all sides, constructed by and large using dry masonry stones with very little mortar. Petrie reported that the stones he observed were an average of 32 inches wide by 58 inches long and 20 inches high.

As each layer is built and rises metre by metre, the weight of the masonry is partly supported by friction, due to the incline on to the core, but mostly down its length, as if it were a vertical member. Pressure would build at the base of the layer proportional to its height and would be even more problematical as the load would be eccentric to the vertical causing a compressive force on the inner face and a tensile force on the outer face. Any sporadic masonry tie would shear and fail under the load, and the layer would slide down to the weakest point of the structure, rotating outwards and collapsing. This would in turn affect the other attached layers, causing a catastrophic failure. This failure, in my view, occurred before the pyramid was completed and would have made completion impossible.

The only mystery is why the centre core did not also collapse if it was constructed in the same way. Petrie again observed added layers at ground level when his party tunnelled into the layers at the base of the structure. My guess is that the tower was not so constructed right to the top of the pyramid. I believe that the visible bands of exposed stonework at the base and point of the collapse, and the smaller band approximately halfway between that and the top of the pyramid are not merely the capping of an accretion or the indentation linking the outer layer into the main structure, but a more substantial form of construction. The stones projecting out from the main body at these positions give the impression that the failure is due to action outside the main tower. If it was a problem with an indentation from the outer layer into the main tower, why are there no voids or indentation into the tower?

This layer of stonework, I think, is a floor that acts like a diaphragm and is built across the pyramid at these levels to provide stability to the central core. The lessons learned must have had a significant effect on the builders.

The essence of engineering is to learn the lessons from your mistakes and failures. The builders did not need to be mathematical geniuses to understand what works in practice and what doesn't. That they understood these faults and addressed them can be seen at the next monument to be built at Dahshur, the Bent Pyramid. They now had some means of providing an adequate open space for the burial chamber inside a massive stone structure. The burial chamber can be above ground level, so that no further excavation is needed for it.

The use of vertical cladding is less stable than interlocking horizontal stone blocks. The core blocks must be of a size and mass to hold the infill within the pyramid together while providing a platform for the outer finishing casing layer.

Armed with new knowledge, the Royal Master Builder together with the pharaoh planned one of the most intriguing pyramids in the evolution of these ever-evolving structures. Despite the previous failure, the Bent Pyramid was constructed initially as a traditional pyramid shape, but again, as with all new technology, the devil is always in the detail.

The Bent Pyramid is located at the royal necropolis of Dahshur, approximately 40 km south of Cairo, built under the Old Kingdom Pharaoh Snefru (2600 BC). It is one of the most interesting pyramids built because it was not completed and it gives archaeologists and engineers an insight into how it was constructed, how the builders again were able to learn from their mistakes and rectify them in the next series of pyramids.

The lower part rises from the desert at an angle of 54°, but just over halfway up the pyramid the angle changes to a much shallower 43°, giving rise to the name Bent Pyramid. Archaeologists now believe that the Bent Pyramid represents a transitional form between step-sided and smooth-sided pyramids. It has been suggested that, on account of the steepness of the original angle of inclination, the structure may have begun to show signs of instability during construction in the burial chambers. This forced the builders to adopt a shallower angle to reduce the overall load on the burial chamber and prevent the chamber from collapsing.

INTERIOR PASSAGES

The Bent Pyramid has two entrances, one fairly low down on the north side, to which a substantial wooden stairway has been built for the convenience of tourists. The second entrance is high up on the west face of the pyramid. Each entrance leads to a chamber with a high corbelled roof: the north entrance leads to one below ground level, the west to a chamber built in the body of the pyramid itself. A hole in the roof of the northern chamber (accessed today by a high and rickety ladder 15 metres long) leads via a rough passage to the corridor from the western entrance. The western entrance passage is blocked by two stone blocks which were not lowered vertically, as in other pyramids, but slid down 45° ramps to block the passage. One of these was lowered in antiquity and a hole has been cut through it, the other remains propped up by a piece of ancient cedar wood. The connecting passage enters between the two portcullises.

External elevations of the Bent Pyramid.

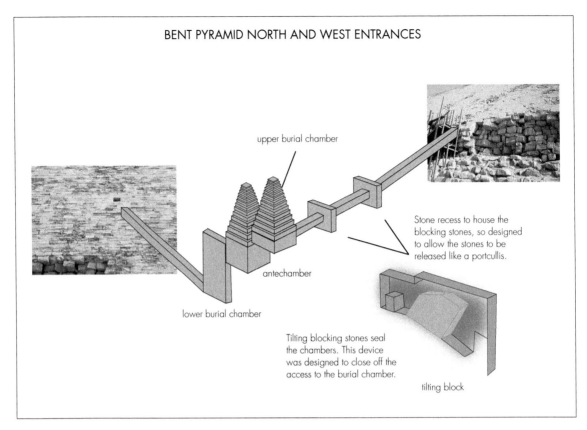

BENT PYRAMID NORTH AND WEST ENTRANCES

upper burial chamber

Stone recess to house the blocking stones, so designed to allow the stones to be released like a portcullis.

antechamber

lower burial chamber

Tilting blocking stones seal the chambers. This device was designed to close off the access to the burial chamber.

tilting block

The interconnecting access to the two burial chambers of the Bent Pyramid.

THE PYRAMID TEMPLE

On the east side of the pyramid there are fragmentary remains of the pyramid temple. Like a pyramid temple of the Meidum Pyramid, there are two stelae behind the temple door; of these only stumps remain. There are no inscriptions or other traces of the structure to be seen. The remains are fragmentary but it is presumed that it was similar to the temple at Meidum.

Returning to the second question of the failure of the outer casing on the Bent Pyramid, I was asked by the authorities to give my opinion on saving the remaining external cladding on the pyramid, particularly, the section at the angle between two edges on the diagonal slopes.

Before any structural restoration work is considered the exact nature of the defects must be established so that the correct intervention can be carried out. From a visual inspection, it was evident that the structure was showing distress along all the extremities. What are the clues? The pyramid does not appear to have any foundation movement. All the missing cladding appears at interfaces or change of direction at the angles and between the ground and the cladding. Could local opportunist thieves have taken this? At the lowest levels, that could be the answer, but at high level and in such a random manner without any sign of indentations for temporary scaffolding or of any symmetrical cutting of the blocks to aid removal, it does not seem likely or even possible. It would have been extremely dangerous work.

Normally, to dismantle a structure, you need as much scaffolding as you would to build it, and opportunist thieves would hardly have had sufficient resources. Indeed, if they merely wanted rough stones they could have had them in the hills adjacent to the centre of Cairo without all the trouble of removing and carting them thirty miles out of town. The damage looks as if it had been caused by a giant sweeping his hand across the face of the pyramid with enormous energy, sucking out the facing and leaving the ragged empty sockets.

In the case of the Bent Pyramid and, I believe, of all the pyramids, the outer casing has been affected by thermal movement. Fortunately, the Bent Pyramid has some degree of stone casing still attached, and the mechanism of failure is clear. The failure of all the perimeter edges shows that the outer casing has expanded from the centre outwards and this movement has taken place on all the extremities. The photographs of the Bent Pyramid show how thermal expansion has caused the blocks to move to the edges where they have detached. They also show how unsupported individual stones can cantilever, snap off and subsequently fall to the ground.

A graphic illustration of the thermal movement at the corner of the Bent Pyramid.

CONDITIONS AFFECTING THE CONSTRUCTION

The next question would be whether the pharaoh specified the size and height of his pyramid or if that was decided by the physical characteristics of the geology of the area where it was to be built.

The western side of the Nile was the start of the Western (Sahara) Desert. It was a sea of sand with hidden islands of stone that would be a suitable base and foundation to carry the enormous load of a pyramid 146 metres high and 230 metres square. The survey team would have called on local knowledge and also their skills in locating a suitable site. I believe this would be their primary task.

The next consideration would be the availability of the core stones and other material needed to construct the pyramid, although, if the core stones were not required to be positioned throughout the pyramid in great numbers, they would not be a primary consideration. All these questions at the start of the project would be subject to what the pharaoh wished and what the Royal Master Builder could achieve within his design and specifications. That is not to say that once the design had been completed and accepted by the pharaoh it could not be changed. It probably was on many occasions through site circumstances and a whole range of other reasons.

Once the project was given the green light, the Royal Master Builder would select and gather his team together.

The logistics of preparing the site, and organising the manpower and materials for a period of up to twenty years must have been enormous and required management and communication skills that were far beyond anything achieved earlier. A permanent site administration office would have been needed close to the actual construction work to control all aspects of the undertaking. All the trusted lieutenants would assemble under the direction of the Royal Builder, the responsibility for each part of the work allocated to each supervisor and then on to the requisite foreman and gang leader. Multiple tasks would be set in motion to prepare for a physical start on site. All of them would have been planned to run concurrently, but with the tasks coordinated so that the project would work seamlessly.

THE METHOD STATEMENT

The Royal Master Builder would create the method statement that would include the main objectives relating to the overall design. This programme of work would provide a detail of when completion of each task would be required, thus providing the back-up support and specialist providers, and stating when they were needed. All this will be very familiar to modern-day project managers without computer assistance.

The method statement for the Great Pyramid laid down how the pyramid was to be built and the order in which every stage of the work was to be undertaken. This was crucial, as it dictated the supply chain for the craftsmen, materials and logistical support.

Having decided the location of the pyramid and its overall size, there now comes the crux, the method of construction. The setting out of the pyramid and its building plan has never to the best of my knowledge been explained. All the textbooks give the overall size of the pyramid and the correct angle of the pyramid viewed at the face elevations, which is correctly stated to be 51.50° on all but the top of the Bent Pyramid and the next pyramid constructed in sequence, the Red Pyramid.

However, the most important elements when setting out any form of construction are the 'cardinal points', the corners and the apex of the pyramid. Without these control points the builder will not be able to establish a balanced construction, and will achieve only a mismatched, out-of-line and defective structure. This was even more important in those early days when builders had no modern surveying instruments or any precise way of measuring angles. They would rely on level, balance and ratio to provide the guidelines.

This method has been handed down to us as builders. In every brick and stone building constructed you see the legacy of this method of construction. The structure is set out at the corners, the cardinal points, which is where the construction starts, using great care and precision to make the courses both level and perpendicular. The construction then rises out of the ground with the brick or masonry filled in between, providing a balanced and level structure until the building is complete. This is a very simple approach when viewing modern construction, but it is the fundamental principle underlying all building work.

The Royal Builder would also be armed with the knowledge of the problems and failures that occurred during the building of the first pyramids. He would be especially aware of the importance of correct corbelling to carry the heavy weights above the entrance corridor to the burial chamber in the Great Pyramid.

He would also have learned from the dramatic collapse at the pyramid at Meidum and would have planned to have a stepped outer core of stones to contain the infill material and also provide a key to attach the outer final casing to the pyramid securing it to the side of the pyramid. But at that time they knew nothing about expansion joints and the damage that would be caused if they were not used.

DUE NORTH AND SETTING OUT

Most of the books, documents and papers on the accuracy of the positioning of the pyramids relate to the north/south aspect of the structures. As a navigator both at sea and in the air, I must pose the question: what would the ancient Egyptians have known of the polar positions of the earth when they believed the world was flat? Yes, they would have a name for north and south, but what importance would they have attached to these cardinal points used in modern navigation and position finding?

I am sure that they studied the heavens in great depth and would have had a good knowledge of both the stars and planets, but would they have had the navigational knowledge to put these to a practical use other than as signposts in the sky.

The biggest influence on their lives would have been the sun. This gave them light and warmth, and by careful observation would have predicted seasons and the time of day. The discovery that the earth was not flat was made by the global circumnavigation of the earth and the first use of a magnetic compass by the intrepid explorers of the Renaissance. Its significance in pointing to magnetic north would not have been known by the ancient Egyptians. As every schoolchild will tell you when asked, the Earth rotates around the Sun approximately once a day, as was discovered by Copernicus in the sixteenth century.

The observed sunrise position will vary depending on the observer's location in terms of latitude, altitude, time zone and season. This is because the earth is tilted

THE PROBLEM OF FINDING ASTRONOMICAL DUE EAST/WEST

Equinox: day and night equal

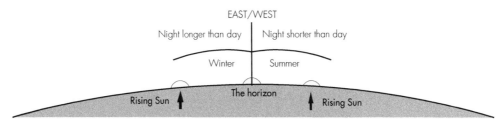

Sun's astronomical observation at sunrise

The sun god Ra was the main god of the ancient Egyptians. He controlled their very existence. He brought light and heat to sustain every facet of their lives. Flooding the Nile each year brought rich, fertile mud necessary to grow their crops and feed their animals. The Sun would have been one of the most studied astronomical heavenly bodies. Its movements would have been well known and documented. The Egyptians divided a normal day into two equal periods of 12 hours. However, there were only two times a year that the period of daylight exactly matched that of the night. These were the Vernal and Autumnal equinoxes, falling on 21 March and 23 September each year. These would mark the position of true EAST/WEST. All pyramids were constructed on a true EAST/WEST orientation and also by definition true NORTH/SOUTH, with a precision that astounds modern archaeologists and engineers.

Suggested methods of setting and marking out a pyramid

A very accurate method of setting out large areas with great accuracy is by the use of transists. When the two control points have been set, the connection between these points can be controlled very accurately by ranging lines of rods or sticks that are placed to achieve a straight line.

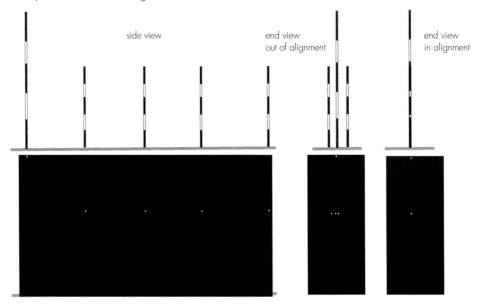

In the desert, both day and night, the obvious way to create straight lines would be to use transists. In the day, these could be ranging poles positioned as illustrated. A series of small lamps could be used at night.

at an angle of 23.5°, and the Sun apparently rises in the east and sets in the west as the Earth rotates around the Sun. This is only a general indication of true east and west, which will depend on the observer's location and the time of year of the observation.

If an observer records the position of the sun as its upper edge touches the horizon, probably with wooden poles in line or transits and records these positions, the apparent movement of sunrise will be observed from the north-east to the south-east quadrants. By observing these he will have established that during March and September at the equinox (when the period of day equalled the period of night) the position would be central. This would indicate a position of true east to west, and the direction would be used as the prime cardinal position of the sun god's movement. By doing this, people discovered north and south in the process.

These observations, I am sure, would be recorded in stone as the base line of the pyramid to be constructed.

Turning to the construction of the Great Pyramid, this would be the first element of the construction phase. The setting out and orientation of the pyramid would be decided by observation of the sunrise at the equinox in March or September.

Setting out the foundations of monuments with such precision, both in shape and level, to an extraordinary degree of accuracy is normally associated with modern-day surveying equipment. The use of water to provide a uniform level has been cited as one of the methods of levelling the base of the pyramid. This would need a channel cut into the stone base or stone blocks created to provide a channel on top of the pyramid base slab. Whichever method was chosen it was also important to have a base datum point level at each corner outside the pyramid perimeter to measure the vertical courses of the stone as each course was erected.

Flinders Petrie and J. Cole had discovered stone sockets at each corner that had no apparent use and did not align with the outer casing or overcladding. These sockets, I believe, were used to erect profiles made from timber. These can be seen on all building sites when the foundations are set out and are used to project the building line to create right angles.

I believe that the ancient builders could also use visual sighting methods such as transits to mark out and measure lengths. The transits would be used to project a straight line, and I would have poles marked in cubits, say 15 to 20 cubits long, to measure the distances and without any rope or cordage. The transits would also be very effective, if used at night with small lights or lamps instead of poles.

Once the base had been set out precisely, the angle of construction would be approximately 45°, which is measured across the diagonals, not the face of the pyramid. A recently published survey by the Glen Dash Foundation together with the Ancient Egypt Research Associates, including previous work by Mark Lehner and David Goodman (1984), in February 2015 provided a detailed, accurate footprint of the Great Pyramid. The difficulty, as always with ancient monuments that have weathered and aged over such a long period, was to establish the base lines. This survey produced new estimates for the size and orientation of the Great Pyramid and confirms the extraordinary accuracy shown by the original builders, confirming the casing lengths as 230.363 metres and the diagonal angle from the base to the top of the pyramid as 44.56°.

OPPOSITE: *How the pyramids could be set out with extreme accuracy.*

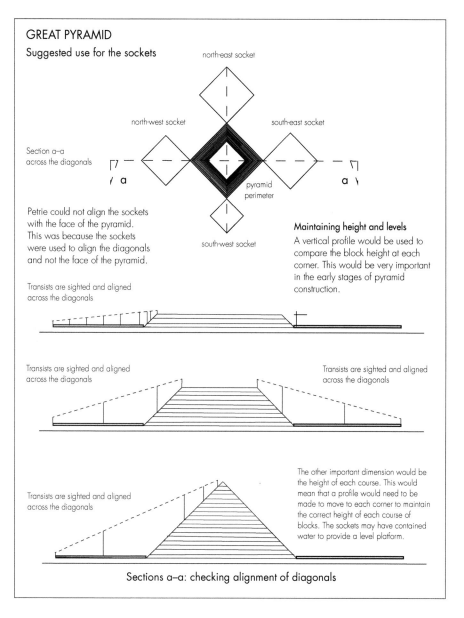

GREAT PYRAMID
Suggested use for the sockets

north-east socket

north-west socket

south-east socket

Section a—a
across the diagonals

a

pyramid
perimeter

a

Petrie could not align the sockets
with the face of the pyramid.
This was because the sockets
were used to align the diagonals
and not the face of the pyramid.

south-west socket

Maintaining height and levels
A vertical profile would be used to
compare the block height at each
corner. This would be very important
in the early stages of pyramid
construction.

Transists are sighted and aligned
across the diagonals

Transists are sighted and aligned
across the diagonals

Transists are sighted and aligned
across the diagonals

Transists are sighted and aligned
across the diagonals

The other important dimension would be
the height of each course. This would
mean that a profile would need to be
made to move to each corner to maintain
the correct height of each course of
blocks. The sockets may have contained
water to provide a level platform.

Sections a—a: checking alignment of diagonals

The angle at the face of the pyramid is a by-product of the pyramid being constructed at the diagonal corners and is not of any building significance. In practice this angle works out slightly less than 45°, measured on the existing structure but that is acceptable as long as all the courses remain level. Thus, the first course of core blocks would be positioned and levelled very accurately with the A-team masons laying the first courses to form the base of the pyramid. The course heights would be measured with vertical storey rods, wooden poles with each course height marked on the rod, so that these lengths can be transferred to each corner to control and maintain the course height.

PYRAMID GEOMETRY
Approximate dimensions and angles

Dimensions updated from the survey
made by Glen Dash for the Ancient
Egypt Research Associates in 2015

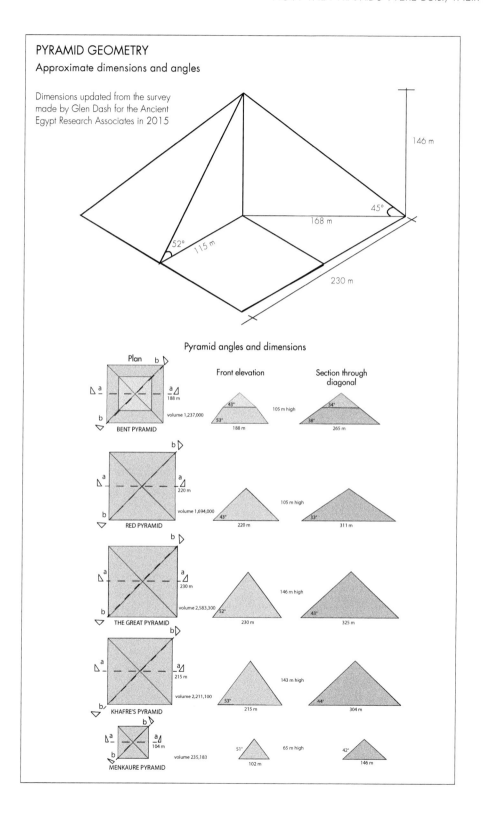

146 m

168 m

45°

52°

115 m

230 m

Pyramid angles and dimensions

Plan

Front elevation

Section through diagonal

b

a — 188 m

volume 1,237,000

BENT PYRAMID

43°
53°
188 m
105 m high

34°
38°
265 m

a
220 m

volume 1,694,000

RED PYRAMID

43°
220 m
105 m high

33°
311 m

a
230 m

volume 2,583,300

THE GREAT PYRAMID

52°
230 m
146 m high

43°
325 m

a
215 m

volume 2,211,100

KHAFRE'S PYRAMID

53°
215 m
143 m high

44°
304 m

a
104 m

volume 235,183

MENKAURE PYRAMID

51°
102 m
65 m high

42°
146 m

The exposed corner of the Bent Pyramid: the first points for construction.

It would be extremely easy for the masons working from the corner at an angle of 45°, which is obviously half of the vertical 90° perpendicular. They would only need to place the next vertical course of blocks at the same distance from the corner that matches the vertical height of the new block to automatically obtain a 45° angle. On the top section of the Bent Pyramid and the next pyramid to be built, the North or Red Pyramid at Dahshur, this angle was 33°. This would be measured by the builders as three units horizontally to two units in height.

The corners of the pyramid would be raised to a convenient height to establish the overall shape and size of the pyramid, with the core blocks being levelled and aligned between the four corners to control the outer perimeter or curtain wall. The thickness of this outer wall would be greater in the corners, even as much as 25 metres, but probably only four blocks thick at the face of the pyramid. This is determined by the angle of the internal ramps when building the pyramid from the inside.

Viewing the front elevation of the pyramid as the super-core blocks are erected at the corners, one would be able to see the advantages of constructing the pyramid from inside out. The pyramid measured 230 metres at each elevation. Excluding the area taken up with the construction of the corners there would have still been close on 200 metres of access into the main body of the pyramid from all four sides. The builders would have used this opportunity to bring in all the heavy large blocks of stone and all the granite for the construction of the burial chambers and the enormous stones for the relieving arches.

This would be the only time that the previously envisaged external ramps would be needed to gain access to the interior of the pyramid. Once the builders were inside the internal perimeter or curtain wall, internal ramps would be constructed to provide working access to the corridors and burial chambers and the movement of any large blocks that were deemed necessary. With this amount of room available, the ramps could be almost 10 metres wide without causing any congestion within the construction area.

Flinders Petrie found that the Great Pyramid had distinct hollowing at the centres of each face in the core stones but not the overcladding stones that were visible at ground level. This would coincide with the break in construction between the corner that allowed the builders access to the central core of the pyramid.

This infill would by its very nature comprise much smaller and more easily managed stones similar to those we found in the centre and above the burial chamber in the Step Pyramid. Having dry-diamond drilled hundreds of metres of stone and examined the stone cores that resulted from this technique, we never found a stone larger than 400 mm across the face. We also had plenty of evidence of much smaller stones bonded together with the tufla grout giving continuity to a stone course.

With this in mind the whole question of the time needed to construct the monument is put into perspective. The idea of impractical large ramps that would have been bigger than the pyramid, or the wrap-around ramps that would invite the same problems, comes under extreme scrutiny and is shown as not fit for purpose.

The method of constructing a pyramid using internal ramps

The builder would have started at each corner after the surveying team had accurately set out the foundation plan.

Outer core casing comprising a few larger stones provides a support for the fill and a key for the outer casing.

The corners would have been built first, with the centre sections left open to allow access to the centre of the pyramid. Internal ramps would be constructed from quarry waste and the burial chamber complex would probably have been built first, with its own supply ramps that would back up the granite and facing stone. The corbelled internal roof would have probably needed scaffolding to position the cantilevered ceilings.

The outer core blocks would be larger at the bottom of the pyramid and would diminish in size as the pyramid grew.

External scaffolding would be used to assist the positioning of the core blocks.

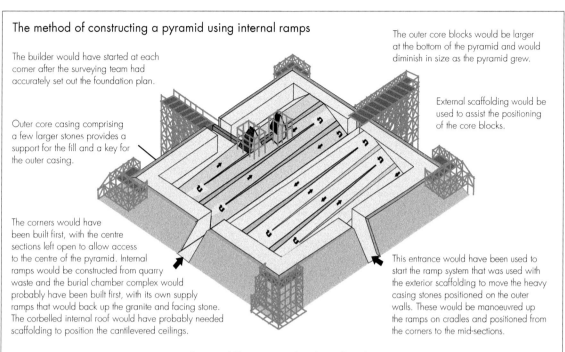

This entrance would have been used to start the ramp system that was used with the exterior scaffolding to move the heavy casing stones positioned on the outer walls. These would be manoeuvred up the ramps on cradles and positioned from the corners to the mid-sections.

Because the internal fill is a mixture of mostly small-sized stones and grout, the ramps could have been relatively narrow, providing gradients well within the capacity of men and donkeys.

Second stage construction

As the outer walls were built, so the scaffolding requirements would have been more extensive. However, the scaffolding could have been limited to the areas under direct construction if timber was in short supply. In any case, the additional outer casing could not be attached without the extensive use of scaffolding, which also meant that any external ramps were unlikely to be used to construct the pyramid.

The ramps would be narrower and steeper as the pyramid was built.

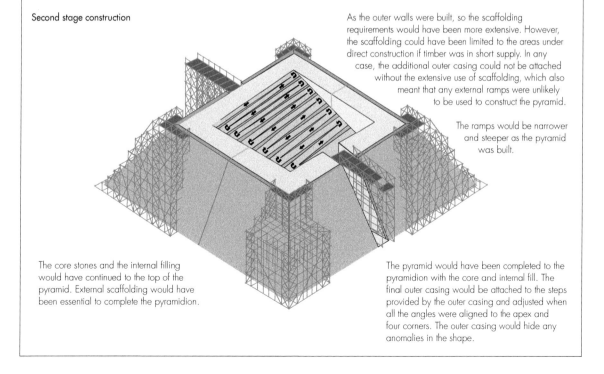

The core stones and the internal filling would have continued to the top of the pyramid. External scaffolding would have been essential to complete the pyramidion.

The pyramid would have been completed to the pyramidion with the core and internal fill. The final outer casing would be attached to the steps provided by the outer casing and adjusted when all the angles were aligned to the apex and four corners. The outer casing would hide any anomalies in the shape.

CONSTRUCTING THE GRAND GALLERY, BURIAL CHAMBER AND RELIEVING ARCHES OF THE GREAT PYRAMID

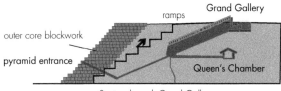

Section through Grand Gallery

The construction of the Grand Gallery, the burial chamber and the relieving arches commences as soon as the pyramid corners are positioned. Utilising the large space between all the corners, ramps are constructed at shallow angles with flat platforms at each change of direction. The large granite blocks are then manoeuvred to the desired height and placed in postion as the structure rises.

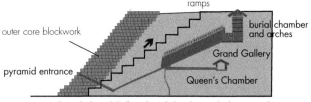

Section through Grand Gallery, burial chamber and relieving arches

The beams are then moved up the next ramp towards the Grand Gallery

The beams are then moved horizontally to start of the new slope direction

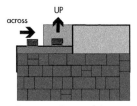

Section at change of direction

Large beams are mounted on sledges and manoeuvred up the shallow inclines until the change of direction

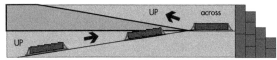

Section at front of ramps

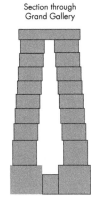

Section through Grand Gallery

The Grand Gallery chamber leads to the burial chamber and is 47 m long by 8.5 m high. The extreme loading imposed by the size and weight of the pyramid on the opening has been compensated by the use of small cantilever corbels.

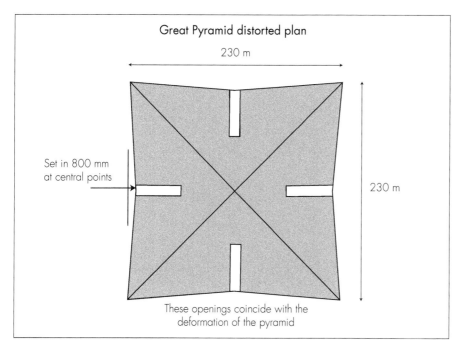

Great Pyramid distorted plan

230 m

230 m

Set in 800 mm at central points →

These openings coincide with the deformation of the pyramid

The distortion in the sides of the Great Pyramid, indicating the access points for the construction.

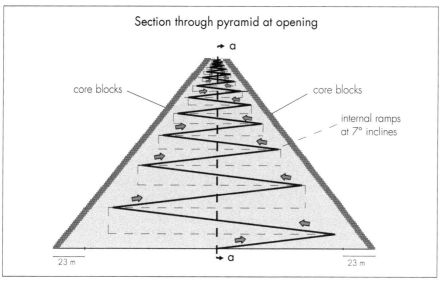

Section through pyramid at opening

→ a

core blocks

core blocks

internal ramps at 7° inclines

23 m

→ a

23 m

Line diagram indicating the feasibility of using internal ramps to build the pyramid.

However, internal ramps would be easily constructed at angles of less than 7°, and probably only 5°, during the initial ramp formation to construct the burial chambers. The direction of the ramps could also be varied to accommodate the construction of the burial chamber complex and provide the method of raising the large core stones now visible from the outside. These would be transported and moved into position along the internal ramps to the perimeter curtain wall and finally positioned with the aid of the external scaffolding that would allow the masons the all-round accessibility needed to position and level each block.

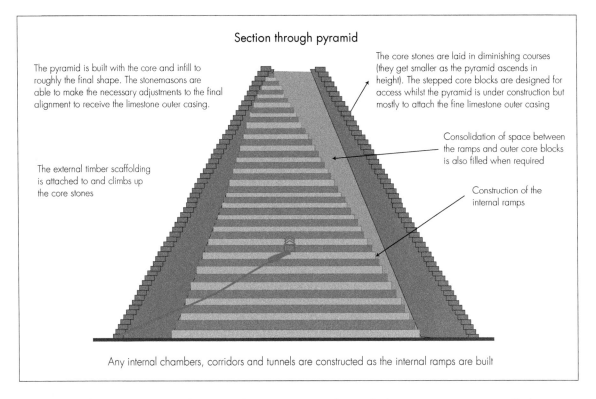

Section through pyramid

The pyramid is built with the core and infill to roughly the final shape. The stonemasons are able to make the necessary adjustments to the final alignment to receive the limestone outer casing.

The core stones are laid in diminishing courses (they get smaller as the pyramid ascends in height). The stepped core blocks are designed for access whilst the pyramid is under construction but mostly to attach the fine limestone outer casing

Consolidation of space between the ramps and outer core blocks is also filled when required

The external timber scaffolding is attached to and climbs up the core stones

Construction of the internal ramps

Any internal chambers, corridors and tunnels are constructed as the internal ramps are built

Cross-section of line diagram showing the internal ramps.

As the internal ramps rise together with the perimeter curtain wall, the ramps would decrease in width to utilize fully the internal space and maintain the ramp angle that would not be more than 12° in pitch. This was why the pyramid was built with diminishing courses. As the pyramid grew in height, so the core stones were reduced in size to cope with the increase in ramp pitch, but were still of sufficient mass to contain the inner fill at each level and to carry the outer case when fixed at a later time.

I have in the past been criticized by archaeologists on the grounds that the Pharaoh Khufu as 'God on Earth' would not have had any infill in his tomb but only full-size core blocks. If this was the case, why did he allow the blocks to reduce in size as the pyramid was built reaching for the sky? The reason is obvious: it was an essential requirement for the construction.

Because of the ample space available at this stage of the construction, the internal ramps could be made in any direction to construct the royal burial chambers, in particular to build the approach to the chambers and the relieving arches above through the corbelled entrance passageway.

Whole palm trees at 90° to the direction of the travel would be available to lever and slide the largest beams and stone slabs incrementally to the burial chamber heights. There would be ample opportunity to have copious amounts of wet clay and tufla grout to assist in sliding the massive stone blocks into position.

Internal scaffolding would also be extensively used to provide 360° access platforms to position the larger stone elements and particularly the Grand Gallery and corbelling.

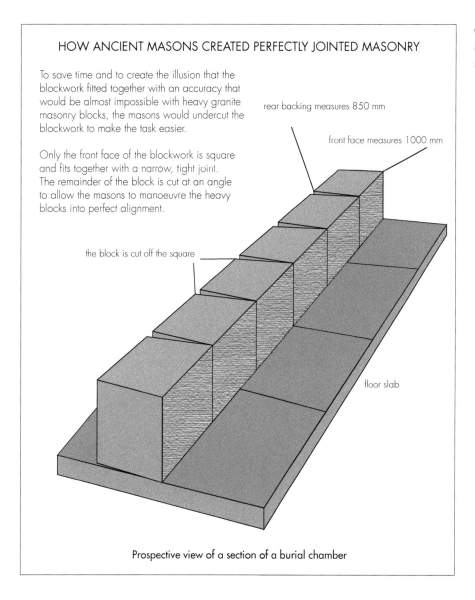

HOW ANCIENT MASONS CREATED PERFECTLY JOINTED MASONRY

To save time and to create the illusion that the blockwork fitted together with an accuracy that would be almost impossible with heavy granite masonry blocks, the masons would undercut the blockwork to make the task easier.

Only the front face of the blockwork is square and fits together with a narrow, tight joint. The remainder of the block is cut at an angle to allow the masons to manoeuvre the heavy blocks into perfect alignment.

rear backing measures 850 mm

front face measures 1000 mm

the block is cut off the square

floor slab

Prospective view of a section of a burial chamber

Cutting the blocks off the square to make very narrow, tight joints.

The granite for the burial chambers would be carefully positioned using a technique known as cutting off the square. This is a method used by masons to give the appearance of very thin joints when the wall is observed from the front face. To create a perfectly square, squared block, 1 m x 1 m and weighing tons, would be a Herculean task. So, the masons would cut the blocks off the square in order to make the joints very thin, making the positioning of the blocks relatively easy.

I have experience of this technique from using it at Stowe, Buckinghamshire. It was used on an external gable end wall of the main building. Because the wall had been exposed to inclement weather over the centuries, it started to bulge and move out as a result of the original undercutting of the stone. Fortunately, we were able to stitch back the outer undercut wall to a newly formed inner wall.

If you look very carefully at the joints in the burial chamber and Grand Gallery you can see where the mason had slipped when chiselling the outer face of the block, revealing a pin prick hole behind the joint created by the undercutting.

It is at this point that I must comment on the large external ramp theories that have been generally put forward as the means of constructing the pyramids. The theory is based on the assumption that the pyramids contain several million large blocks, such as seen on the external elevations that continue through the entire pyramid at each level.

Let us start with the block theory which states that the pyramid is full of squared and dressed stones similar to the outer core faces. This gives rise to a number of impossible claims. The first is that the time it took to construct the pyramid would have been extended by at least another quarter over all. Anyone who has tried to fit a number of children's wooden blocks into a known shape such as a box will know that placing squared blocks, even ones that are all the same height into a confined area or perimeter, takes much longer because they have to be carefully fitted. Now consider that this has to be done at each course level with varying block heights within the pyramid.

The long-held ramp theory dispelled.

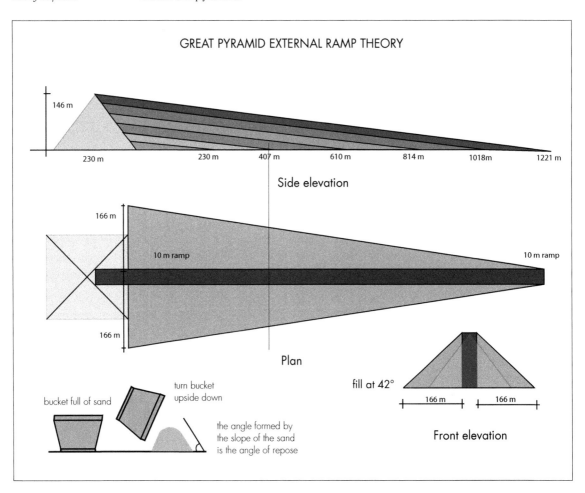

Now let us look at the quarry close to the Great Pyramid that supplied the core and internal stones; it has been calculated to hold 2,700,000 cubic metres of stone. Excluding a very small percentage of granite used in the burial chamber, this amount would not be nearly sufficient to supply the need.

Having located a suitable source of stone, the top covering of sand and debris would need to be removed to reveal the selected stone. Next it is necessary to locate, mark out and cut 2,300,000 blocks and dress them to fit each course; then to transport them on paved roadways to the pyramid or the start of the external ramp previously mentioned; then to move them individually and without mishap up a ramp that has to be continually lengthened as the pyramid gains height to its final level, to be placed in position not just at that point where it is at its correct height but to be manhandled and mortared into position metres away at the other side of the pyramid.

It has been calculated that this method of construction would need to have been undertaken every six minutes during a ten-hour day for twenty-five years to complete the construction of the pyramid. This also leads to the question of how many men were needed to build the pyramids. Surely, this method would need the 20,000–30,000 mentioned in many of the articles published to date. There has been much speculation regarding these numbers as to whether they were full-time workers or seasonal agricultural workers. The numbers seem staggering and it would be difficult to imagine how all these labourers would be able to work on such a confined site.

The process of quarrying would not be able to supply the number and quality of stones using this method, even if they were available within this time frame. The stones would have to be cut from the quarry face. Therefore, access would be the limiting factor, not the number of workmen allocated to the task. The quarrymen would get in each other's way. A face would have to be worked and totally removed before access to the next face was exposed.

Once the burial chamber had been completed the rest of the pyramid would continue as before. The ramp would continue to climb with the external core perimeter wall, but the angle and the width of the ramp would diminish. This would coincide with a reduction in size of the core stones to make it easier to move into place.

As the ramps gained height the smaller particle sizes would be consolidated as, in effect, the ramps were built layer upon layer, consolidating and vibrating the stone into place. The infill material would be whatever the builders could use, and obviously the larger blocks would be positioned at the base of the pyramid to fill the space and provide additional support.

It was more than possible to use scaffolding to construct the pyramids as can be seen in the photograph showing the current work being undertaken on the Step Pyramid with only traditional timber and rope lashings to replace the damaged outer facade. I believe that this would have been the preferred method of building the pyramids together with internal rather than external ramps, as has been suggested by many others.

Another indicator is a number of rectangular sockets set at the bottom casting blocks of the pyramids of Khufu and Khafre at Giza. The sockets are formed at regular intervals and of varying sizes and depths, in a straight line or staggered. These were undoubtedly used as sockets to receive the first large upright standard legs of the scaffolding used when a high level of loading was expected, so preventing any sideways movement of the scaffold under large loads.

The method of building the pyramid would have been to construct the inner fill initially with large blocks to consolidate the foundation area slightly higher than the outer core walls. The blocks for the outer core walls would be raised via internal ramps constructed of small stones and any rubbish the builders wanted to hide that could be used effectively. The outer core walls would have a greater thickness at the base (say 10 blocks wide) and slightly less as the structure narrows at the apex to a single block, and these would have been built concurrently with the inner core, but slightly lower to facilitate the placement of the larger blocks that would have been moved up the traversing roadways via one of the centrally formed entrances. The ramps would not need to be very wide and could have had small palm trees embedded into the surface to provide a sliding mechanism to assist the craftsmen transporting and placing the stones.

As the pyramid extended in height the outer core stones were reduced in size and the ascending ramps adjusted to the angle and size necessary to transport the core blocks.

The use of traditional rope and timber scaffolding on the repairs to the Step Pyramid.

Attempts were made in the eighteenth century to conduct archaeological treasure hunting with gunpowder at the base of Khufu's Pyramid. The gash exposed a jumble of large blocks that were neither coursed nor aligned. This would lend support to my opinion that the internal fill is a mixture of material and not the block work that is visible on the outer core.

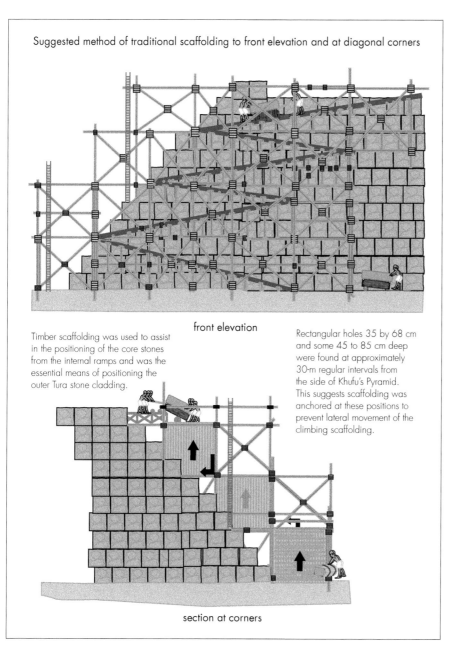

Suggested method of traditional scaffolding to front elevation and at diagonal corners

Suggested method of providing scaffolding to the outer casing of the pyramids.

front elevation

Timber scaffolding was used to assist in the positioning of the core stones from the internal ramps and was the essential means of positioning the outer Tura stone cladding.

Rectangular holes 35 by 68 cm and some 45 to 85 cm deep were found at approximately 30-m regular intervals from the side of Khufu's Pyramid. This suggests scaffolding was anchored at these positions to prevent lateral movement of the climbing scaffolding.

section at corners

There is an easy way to prove this theory, and I would volunteer to carry out the work at no charge to the Egyptian authorities. We could diamond-drill 100-mm core holes into the pyramid at varying heights to a depth of 20 to 30 metres and provide a drilling log of all the contents of the bored hole to establish the true nature of the fill. The drilling would be done with the latest dry-drilling techniques to prevent damage to the pyramid, and the core plugged and filled to match the external appearance. The entire intervention could be catalogued using a drain camera to

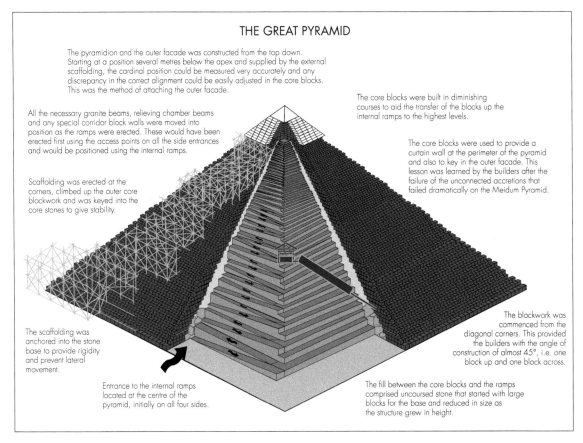

THE GREAT PYRAMID

The pyramidion and the outer facade was constructed from the top down. Starting at a position several metres below the apex and supplied by the external scaffolding, the cardinal position could be measured very accurately and any discrepancy in the correct alignment could be easily adjusted in the core blocks. This was the method of attaching the outer facade.

The core blocks were built in diminishing courses to aid the transfer of the blocks up the internal ramps to the highest levels.

All the necessary granite beams, relieving chamber beams and any special corridor block walls were moved into position as the ramps were erected. These would have been erected first using the access points on all the side entrances and would be positioned using the internal ramps.

The core blocks were used to provide a curtain wall at the perimeter of the pyramid and also to key in the outer facade. This lesson was learned by the builders after the failure of the unconnected accretions that failed dramatically on the Meidum Pyramid.

Scaffolding was erected at the corners, climbed up the outer core blockwork and was keyed into the core stones to give stability.

The blockwork was commenced from the diagonal corners. This provided the builders with the angle of construction of almost 45°, i.e. one block up and one block across.

The scaffolding was anchored into the stone base to provide rigidity and prevent lateral movement.

Entrance to the internal ramps located at the centre of the pyramid, initially on all four sides.

The fill between the core blocks and the ramps comprised uncoursed stone that started with large blocks for the base and reduced in size as the structure grew in height.

Sectional isometric showing the method of constructing the Great Pyramid.

record the full length of the drilled hole. The hole would be made good using a suitable lime grout.

There has been some speculation that there was insufficient timber available to scaffold the entire pyramid. Further, would the builders have the skill to erect a scaffold to the heights needed to build the pyramids? The availability of scaffold over such a long construction period would in my view not have been a problem, particularly if the scaffold was reusable and could be built up in quantities and initially moved to each section under construction, as opposed to scaffolding for the whole pyramid. The scaffolding could also be transferred to the next pyramid under construction even though some of it would have to be replaced. The skills needed to create large runs of scaffolding would be easily acquired, as at that time people were already able to build ships, barges and river craft, and the skill level would have easily been transferred.

If the biggest task was to create the canals and waterways used to provide the essential materials for the pyramid, this would have been the first activity. The construction of dwellings to house the builders, and provision for fresh water and food would start and increase in line with the number of workers employed.

All the necessary trades needed would be recruited – notably scribes, quarrymen, masons, carpenters, scaffolders, blacksmiths, butchers, bakers, cooks, farmers, stockmen, potters, candle makers, tailors, labourers, and probably women engaged in the 'oldest profession in the world'.

The quarrying of stones would start at the nearest available site. The yield of usable good core stones from the quarry could not be guaranteed, particularly as they had to be sized and dressed to provide continuity of shape and size to match the masonry courses. As the stones were laid in diminishing courses as the pyramid grew in height, this problem was not so critical and the transporting and positioning of the blocks became correspondingly easier. The overburden and the waste material would have been considerable and was used initially for road construction leading to the site and later on as selected infill to provide internal ramps to construct the pyramid.

The site preparation would also provide some large infill stones particularly where a stone outcrop within the pyramid base needed to be reduced in size or removed for access reasons. Excavated material from subterranean chambers would also be used as fill.

Logic would say that the bigger blocks of infill stone would be used at the base of the pyramid and especially at the corners inside the core stones. The core stone, having been cut from the quarry, would have to be dressed, sorted into usable sizes and transported to a location next to the pyramid ready for use. The wooden transporting sledges would have been designed to hold a variety of sizes and shapes that could be pulled either by men or by oxen, to suit the weight and size of the load. Smaller stones for infill could again have been put onto sledges or carried by men or donkeys.

There must have been a large collection area for the storage and final dressing of the core stone very close to the construction. Masons could gauge the stones for each course and lay them as required into their final position.

White Tura limestone outer casing to complete the pyramid would be the last development for the builders. Once a source of supply had been established during the construction phase of the pyramid, this would have been quarried in suitable modular shapes and transported, probably by barges, to an assembly area adjacent to the pyramid site. Again, the stones would have been graded for size and suitability and stored in rows for the masons to use course by course.

The construction of the outer facing would not have commenced at the bottom of the pyramid but at the very top. The cardinal rule applies. The four corners and the top are the builder's guides, and these are the golden rules for construction of the pyramid. Get these wrong and you are in big trouble.

The distinguished French Egyptologist Gaston Maspero, who succeeded Auguste Mariette as director of the Egyptian Museum at Bulaq, worked with Flinders Petrie on numerous sites including Saqqara and first published the pyramid texts, also believed that the outer casing was attached from the top down. However, he believed that this was to conceal the pyramid entrances from spoilers and robbers and not as a technical construction requirement.

The pyramidion would have been constructed in sections and hoisted via an external scaffold to a platform erected at (say) 5 metres below the apex. This platform would have been erected on all four faces at the same level and tied into the core stones to provide a solid base. The pyramidion would have been positioned and grouted together, and any dressing or shaping would have been done at that time. The master masons would have been able to use this fixed cardinal point to

THE GREAT PYRAMID

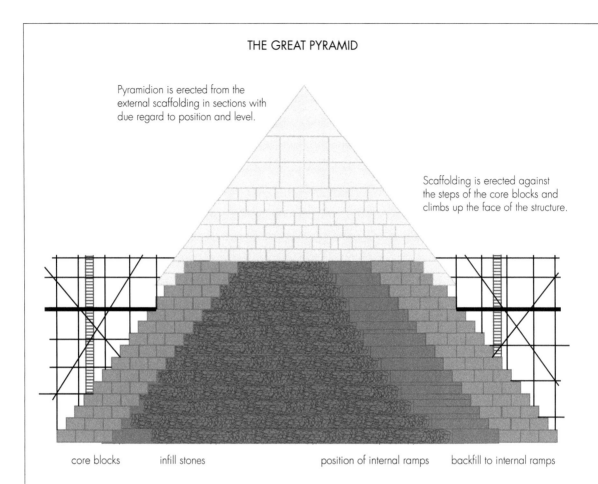

Pyramidion is erected from the external scaffolding in sections with due regard to position and level.

Scaffolding is erected against the steps of the core blocks and climbs up the face of the structure.

core blocks   infill stones   position of internal ramps   backfill to internal ramps

Section through pyramid

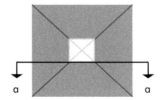

a   a

The core blocks are built in diminishing courses from the base to the apex of the pyramid to facilitate the internal ramps and external scaffolding.

The outer white Tura cladding is positioned in sections from the top of the pyramid to the base, keeping all angles and levels in alignment.

The internal ramps are filled in after they have served their use.

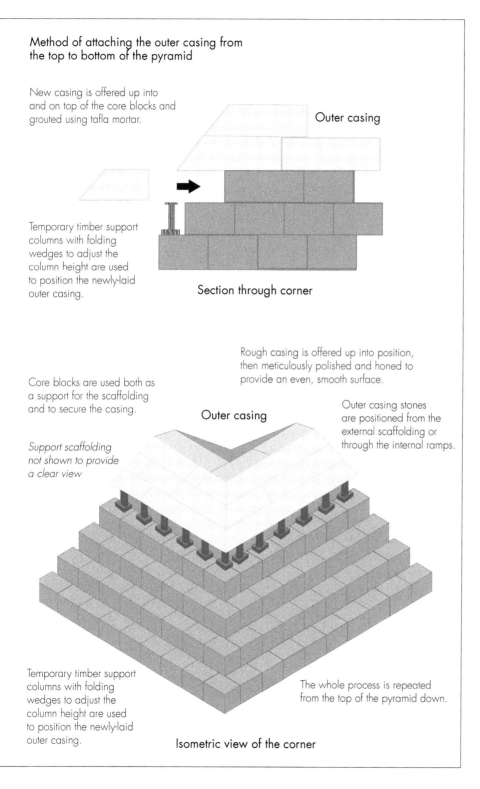

Method of attaching the outer casing from the top to bottom of the pyramid

Constructing the pyramidion and outer casing from the top down.

New casing is offered up into and on top of the core blocks and grouted using tafla mortar.

Outer casing

Temporary timber support columns with folding wedges to adjust the column height are used to position the newly-laid outer casing.

Section through corner

Rough casing is offered up into position, then meticulously polished and honed to provide an even, smooth surface.

Core blocks are used both as a support for the scaffolding and to secure the casing.

Outer casing

Support scaffolding not shown to provide a clear view

Outer casing stones are positioned from the external scaffolding or through the internal ramps.

Temporary timber support columns with folding wedges to adjust the column height are used to position the newly-laid outer casing.

The whole process is repeated from the top of the pyramid down.

Isometric view of the corner

build the outer casing, using the same technique from top to bottom. Any variation in the line or level of the core stones could be addressed without affecting the overall pyramid shape. The outer case would continue down to the base of the pyramid from top to bottom with the scaffolding being lowered by several courses at a time. Adding or removing the core stone could easily rectify any large realignment when necessary as the problem arose.

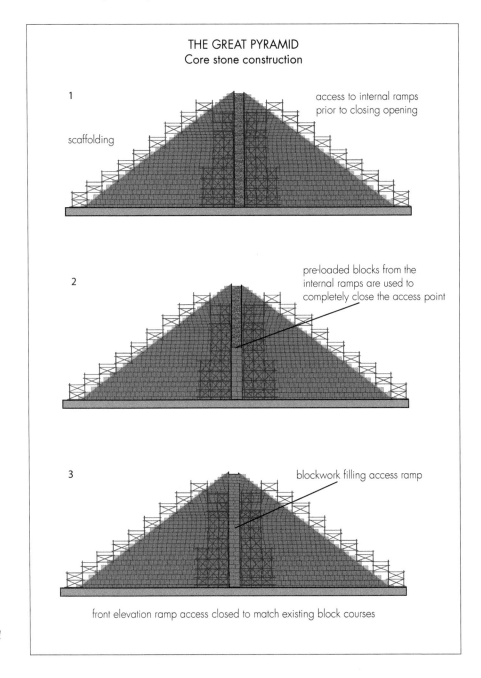

THE GREAT PYRAMID
Core stone construction

1   access to internal ramps
    prior to closing opening

scaffolding

2   pre-loaded blocks from the
    internal ramps are used to
    completely close the access point

3   blockwork filling access ramp

front elevation ramp access closed to match existing block courses

Closing the access openings to the internal ramps.

The very bottom of the pyramid would not need to be squared off all around the pyramid. This has always been a puzzle to the archaeologists who have pondered upon this point.

This technique was used by me when we supplied an anchoring system that included a stainless steel shelf angle at a site in Paisley in Scotland in the late 1980s. The contract was to reattach the outside brick skin to a fifteen-storey high-rise building that had failed. The problem was that a new wall had to replace the failed existing wall. Alignment of the new brickwork was critical. It had to connect very accurately to the tower block's existing roof. However, there was some leeway dimensionally at the base of the tower block. The new wall was constructed on cantilevered angle brackets attached to our corbel anchors, and the wall was constructed from the fifteenth floor down to ground level with any dimensional variation adjusted at each floor level, resulting in a perfectly flush new external brick wall.

THE FAILURE OF THE OUTER COVERING ON THE PYRAMIDS

The final failure of the magnificent structures could never have been predicted by the pharaoh or his team of builders. They would not have been able to understand that after they had spent so much time and effort in constructing these wonderful structures to the god Ra, their sun god, he would have been the instrument to destroy the outer casing and undo all their good work.

Returning to the question of the failure of the outer casing on the Bent Pyramid: I was asked by the authorities to give my opinion on saving the remaining external cladding on the pyramid.

From a visual inspection, the structure was showing distress along all the extremities. What are the clues? The pyramid does not appear to have any foundation movement. All the missing cladding appears at interfaces or change of direction at the angles and between the ground and the cladding. Could opportunist local thieves have taken this? At the lowest levels that could be the answer, but at high levels and in such a random manner without any sign or indentations of temporary scaffolding or of any symmetrical cutting of the blocks to aid removal, it does not seem likely or possible. It would have been extremely dangerous work. In the case of the Bent Pyramid and, I believe, all the pyramids, the outer casing has been affected by thermal movement. Fortunately, the Bent Pyramid is the only one with any degree of stone casing still attached, and the mechanism of failure is apparent. The failure of all the perimeter edges shows that the outer casing has expanded from the centre outwards, and movement has taken place on all the extremities.

The photographs of the Bent Pyramid show how thermal expansion has caused the blocks to move to the edges, where they have detached. They also show how an individual unsupported stone can cantilever and snap off, and subsequently fall to the ground. Limestone has a coefficient of thermal expansion of 8×10^{-6} (0.000,001) proportional to the change of temperature and to the original dimensions, and many natural stones retain a minute proportion of the expansion when they cool down and do not return to their original size. If we consider that in the daytime the temperature rises to 40° across the face of the outer casing and then at night cools to 3° because of the lack of cover and exposure to the prevailing winds, this would give an average temperature fluctuation of 37°. Obviously, this varies throughout the seasons, but to illustrate my point I will build these into the calculations. Let us say that $8 \times 10^{-6} \times (40-3) \times 100$ m = 0.0296 m of movement per 100 metres run in all directions. However, this is also dependent on the size of the gaps between each stone and the next. The outer casing would expand to about 40° in the day and cool to about 3° at night. All the movement would be taken up initially in the joints, but the limestone does not revert to its original position. This expansion will create dust and stone particles that would detach from the stones, filling the gaps between the stones. This would have the effect of reducing the amount of contraction at night. In addition to the natural resistance to revert to its original size, the whole cycle will start again. Multiply the number of days the pyramid has been erected by this endless movement and you have the reason why all the outer casing has moved to the extremities where it has buckled or been displaced against other blocks moving in the opposite direction, and then has fallen off and probably only then been picked up by opportunists and removed from site. I believe that this is the mechanism of failure at this and all the other pyramids.

BELOW: *Damage caused by thermal expansion to the extremities of the Bent Pyramid.*

RIGHT: *An elevation of the Bent Pyramid indicating how the thermal expansion detached the outer casing.*

BELOW LEFT: *The only remaining hanging corner of the pyramid that is near failure.*
BELOW RIGHT: *Front view of the missing blockwork.*

BELOW LEFT: *The point of failure captured on camera. The compression of the outer casing causes the block to cantilever and snap off the base of the pyramid.*
BELOW CENTRE: *Lower entrance to the pyramid.*
BELOW RIGHT: *The size of the blocks illustrated in relation to the height of a man.*

ABOVE: *The front elevation of the hanging corner blockwork.*

ABOVE: *The damage at ground level due to the compression of the blockwork at base of the pyramid.*

The missing blockwork added back at the damaged corners using a drawing programme. This shows how the pyramid would have looked when built.

The gradual failure of the blockwork is represented.

The gradual failure of the blockwork is represented.

The gradual failure of the blockwork is represented.

The gradual failure of the blockwork is represented.

The corner detail as seen in 2006.

I have read in a paper by John A. R. Legan based on an article published in the *Göttinger Miszellen*, 116 (1990), 65–72, that the original dimensions recorded by Flinders Petrie were inaccurate and that the dimension taken in 2004 was larger by a small degree. Assuming these measurements were from the same positions as Petrie's, this is what I would expect of a structure that is still moving and increasing in size. In addition, the stones arching between fixed points could cause the convex shape of the outer casing. The transit of the sun across the region will vary throughout the seasons and will have the effect of heating one side more than another, giving rise to disproportionate movement, particularly at the extremities.

Why does the Bent Pyramid still have half of its outer casing attached and the Red Pyramid and the Great Pyramid on the Giza plateau have virtually none?

I believe it was due to the increased skills of the craftsmen, who developed more knowledge and precision as the pyramid construction developed. They were able to provide better accuracy and building quality in jointing the slabs. Probably the Bent Pyramid was built with less care and with more voids between the stones, which acted like expansion joints, and the casing blocks being inclined inwards at the base of the pyramid may have had the effect of limiting the expansion.

6 THE STEP PYRAMID

I WAS ENQUIRING ABOUT the state of play in Saqqara with Engineer Emad of Intro Trading in 2006 when he mentioned problems with the burial chamber in the adjacent Step Pyramid, also at Saqqara and a short distance from the Serapeum. Apparently, there was great concern on how to overcome the problem of the dangerous ceiling at the top of the underground burial chamber in the centre of the pyramid. After a few telephone calls, he had arranged for us to visit the site the very next day to inspect the damaged ceiling.

Located north-west of Memphis on the west bank of the Nile, the Step Pyramid stands erect and dominates the landscape of this immense necropolis of the ancient Egyptians, some thirty kilometres from the centre of the present modern city of Cairo. One could easily be forgiven for calling the pyramid the stairway to the stars because of its appearance and the ancient Egyptians' belief in the resurrection in the afterlife. On the pyramid, most of the outer casing of Tura stone has gone and is replaced with sand blown across from the Sahara Desert. In some places the core masonry also has disappeared from the side of the pyramid leaving gaping holes.

The Step Pyramid measures 121 metres by 109 metres at its base and is 62 metres high. It is recognized as the oldest stone building known to man. It was built for the burial of the Pharaoh Djoser by his vizier Imhotep during the twenty-seventh century BC. Imhotep was the pharaoh's overseer and was regarded as the inventor of dressed stone. He was a mystical man of many talents, including medicine, and the Greeks considered him a god of wisdom.

From its external appearance, the pyramid looks like a massive stone edifice built to resemble a modern-day wedding cake. But on investigation, and by consulting the many drawings made by the French architect Jean-Philippe Lauer, who made it his life's work to prepare detailed drawings of the pyramid, it can be seen that the original step was based on a traditional mastaba, a traditional term for a single-storey tomb, approximately 10 metres high that preceded the pyramids. This was a low mud-brick structure, resembling a bench and consisting of a burial shaft

The Step Pyramid at the commencement of the repairs.

in the centre that descends to the burial chamber and the sarcophagus. At ground-floor level would be a number of rooms that would depict the life of the deceased and a provisions store for the deceased in the afterlife.

According to Lauer, who had spent eighteen years measuring, drawing and detailing the pyramid, it had initially been constructed as a mastaba. However, not content with this arrangement, the pharaoh instructed the overseer to increase the height of the mastaba by another three layers, partly on top of the existing mastaba and partly by extending the base width in keeping with the overall appearance of the pyramid. Later on, another three layers were added, again extending the base width to stabilise the pyramid shape that we see at present. There has been some speculation on why this had happened. One theory is that as the pharaoh's reign was unbroken and he was in good health, additional extensions were made to celebrate his achievements.

Below ground, there is a network of tunnels and passageways leading from the outside of the pyramid towards the burial chambers. This involves approximately 5 kilometres of tunnels, of which the builders were responsible for half, and the other half were dug by tomb robbers searching for buried valuables.

The burial chamber was surrounded by four galleries, which probably contained the funerary equipment of the pharaoh. Some of these were decorated with blue faience tiles, in keeping with his status.

Djoser, also known by his Horus name Netjerykhet (Divine Body), reigned during the Egyptian Old Kingdom period of the Third Dynasty, around 2670 BC (the authorities differ on the dates and length of his reign). His statue was found by excavators in the tomb chamber and was damaged but still intact, and can be seen in the Cairo Museum. However only bone fragments were found, but these could not be confirmed as coming from the mummy.

The high-level entrance into the burial chamber.

The Step Pyramid is an amazing monument that takes man out of the ground and into high-rise structures. It marks the dawn of civil engineering and its construction, together with the other great pyramids that followed in the pyramid construction era, is the forerunner of modern engineering. The lessons learned in this multi-storey structure have been handed down through the ages and copied by modern builders.

Not only was the achievement in constructing the pyramid breath-taking, but when one considers it was built some 4,700 years ago, one can only imagine the logistics required from day one of the project to provide all the necessary labour, equipment, scaffolding, quarrying and cutting of stone, transporting to site, feeding and housing of the workforce and looking after their welfare. This sort of structure could not be constructed by pressed men or slaves. It must have been completed by dedicated tradesmen with artisan skills drawn from all over Egypt.

The ruined mortuary temple is situated north of the pyramid and is the access to the tunnel complexes that run beneath the sarcophagus and the entrance to the middle opening in the burial chamber. The southern tomb almost mirrors the main burial chamber in depth, but its width is much smaller and the entrance is just an opening in the ground with what looks like a vertical drop to the base 28 metres from the surface. Its galleries are decorated with blue faience tiles and three false doors made of stone showing the pharaoh performing a ritual walk as part of the Heb-Sed ceremony held, according to tradition, after thirty years of his reign to rejuvenate him and to prove that he was still capable of ruling.

Beneath the pyramid the tunnels and galleries connect to the burial chamber's central shaft, which was nearly 8 metres square and 29 metres deep when first con-structed. The galleries and storeroom spaces at the bottom of the shaft provide storage for the food and provisions needed in the afterlife. Eleven shafts were built on the east facade of the tomb and were intended for the burial of the pharaoh's wives and children.

I am aware of three tunnel entrances into the burial chamber. The first and main entrance is situated just below ground level on the north side of the pyramid and terminates at the top of the burial chamber some 28 metres from the external face of the pyramid. The middle entrance and one at the bottom of the chamber are both accessed from the tunnels. The lowest access opens directly beneath the sarcophagus.

The first view of the remaining random blockwork of the burial chamber in 2006.

The entrance to the 28-metre shaft was built on the north side of the pyramid, as was commonly done during the Old Kingdom. The underground passages are cut and shaped through the limestone base by the builders, twisting and turning, presumably following the easiest line of resistance. However, one can imagine the problems of navigation and accuracy during this undertaking. The chamber walls are decorated showing the king in the Heb-Sed ceremony.

Lauer's sectional drawing indicates that the burial chamber was constructed from large granite blocks, which he believed were finished by cladding the walls with decorated alabaster panels. Robbers had probably removed all the alabaster cladding in antiquity. Another interesting find was a carved five-pointed star in limestone attached to the ceiling, which lends support to the theory that a fixed star representing pharaohs in the next life can be seen every night when the stars are visible.

When I visited the Step Pyramid in November 2006 with engineer Emad, I knew that the project was in the planning stage and no actual work had begun on the burial chamber.

Detailed investigations were being carried out by the Supreme Council of Antiquities, which included providing temporary lighting through the high-level entrance shaft some 28 metres from the face of the pyramid into the burial chamber. At that time supporting columns made of stone were visible at regular intervals in the centre of the corridor and they were showing signs of distress. Above the main entrance to the corridor can be seen a very large vertical crack that continues through the pyramid bisecting the burial chamber ceiling. Passing along the corridor with head bowed because of the low ceiling, after some 28 metres, we reach the vertical edge with a sheer drop of 29 metres to the base of the chamber and the location of the sarcophagus. In the limited half-light that filtered across the diagonal of the 8-metre square burial chamber, could be seen the ravaged inverted cup of hanging stone.

I estimated that the amount of stone that had collapsed and fallen twenty-nine metres to the base of the shaft since the pyramid was originally constructed, landing on the sarcophagus and damaging it, was approximately five metres to the highest point of the eroded failure.

At that level, it was not possible to gauge the individual size and the magnitude of the myriad stones that were jammed and locked together. The oblique angle of sight, in the poor light painted a breath-taking and fearsome view of a random ripple of stones trying to pass through a funnel that formed this jigsaw of inter-locked stones 8 metres square with a potential height of sixty-two metres above it. One particular stone, weighing approximately a ton, was balanced precariously on a wooden plank that had been used in previous abortive attempts to shore up the vaulted ceiling. I hoped that that stone didn't have my name written on its under-side. Curiously, when we eventually safely disposed of it, it was discovered that the wooden plank was a coffin lid from antiquity with hieroglyphs on the inner side giving the deceased the instructions from the Book of the Dead on how to behave in the afterlife. I hoped that this was not an omen for the future. This would be the most difficult and dangerous project I had ever attempted.

TOP: *A coffin lid used to shore up the ceiling in antiquity.*
BOTTOM: *Attempts to repair the ceiling in antiquity indicating the problems of such a task.*

Looking straight down into the black gloom towards the sarcophagus from the vertical precipice at the junction between the corridor and the edge of the burial chamber, it was impossible to see the base. This was because the limited illumination from our small hand-held torch could only just light up the middle entrance on the opposite side and halfway down the burial chamber. This view was not improved at a later date when we were able to photograph the ceiling from the base of the chamber looking directly up some 29 metres. The view at this range was just of a delicate, docile stone jigsaw of brown hues that gave the impression of a patchwork quilt.

My initial thought was: how could we span and position a cantilever scaffold platform from the corridor entrance into the main body of the burial chamber to effect a repair. What probably triggered this line of thought was examining the repairs attempted in antiquity. The workmen then had tried to arrest the collapse using palm trees and timber baulks to prop up and prevent the stone from falling onto the sarcophagus. I was intrigued about how they achieved this feat without using a scaffold from the base of the chamber.

This was never going to be successful because they had not resolved the initial problem of why the ceiling had originally collapsed. The repair was always going to be a Sisyphean task, like using a sticking plaster to hold back a ruptured artery. This was further accelerated with seismic

ABOVE: *An old photograph taken from the bottom of the chamber showing the attempt to shore up the ceiling.*
RIGHT: *Section of a detail on providing a cantilever option to strengthen the burial chamber ceiling.*

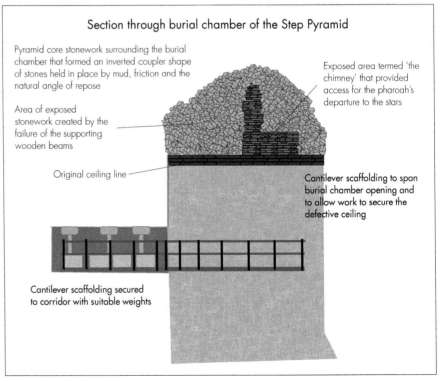

Section through burial chamber of the Step Pyramid

Pyramid core stonework surrounding the burial chamber that formed an inverted coupler shape of stones held in place by mud, friction and the natural angle of repose

Area of exposed stonework created by the failure of the supporting wooden beams

Original ceiling line

Exposed area termed 'the chimney' that provided access for the pharoah's departure to the stars

Cantilever scaffolding to span burial chamber opening and to allow work to secure the defective ceiling

Cantilever scaffolding secured to corridor with suitable weights

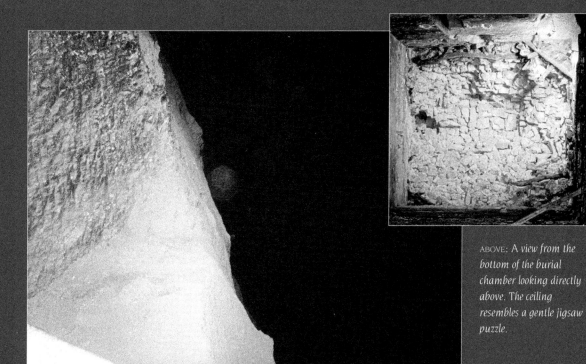

ABOVE: A view from the bottom of the burial chamber looking directly above. The ceiling resembles a gentle jigsaw puzzle.

ABOVE: First view looking down from the top entrance into the jet black burial chamber.

RIGHT: A view from the top of the middle entrance of the burial chamber.

activity, especially during the 1992 earthquake, and the fractured ceiling was continuing to haemorrhage one or two stones every month. The only way to bond the stones was by friction and a dried mud matrix.

Whatever temporary timber supports were used for the last repair came crashing down onto the sarcophagus together with approximately 150 tons of stone. However, they did leave some evidence of the repairs that had been attempted in the past, with scars, holes and some light timber still in place. It was obvious that any permanent repairs could only be achieved if a steel scaffold could be constructed from the base of the chamber with a working platform at the position of the original flat ceiling.

It was then necessary to establish why the ceiling had collapsed. Was it seismic action over the ages that continually eroded the ceiling? Was some other mechanism responsible for the failure? The very detailed drawings by Lauer were a good indication of the construction of the chamber, as was his detail showing the collapsing arched stone at the very top of the burial chamber ceiling, which he had observed in the 1930s.

The Egyptian authorities were also able to say that, as far as they knew, the chamber had been closed for over a century owing to the defective ceiling.

The main evidence of why the ceiling had failed was only discovered when the scaffold working platform had been erected and I was able to gain access to view the remains of the original timber beams that were still in place after 4,700 years. These beams, made from palm trees, had disappeared in antiquity, but one at the very perimeter of the pyramid was still held in place and supported both sides of the burial chamber. The palm tree beam, approximately 300 mm in diameter, was originally over 8 m long and was used to support stonework above the burial chamber. At the time the pyramid was built it was supposed that builders did not have the knowledge or technology to provide stone beams to span 8 metres. They

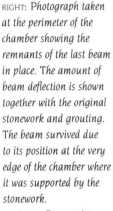

RIGHT: *Photograph taken at the perimeter of the chamber showing the remnants of the last beam in place. The amount of beam deflection is shown together with the original stonework and grouting. The beam survived due to its position at the very edge of the chamber where it was supported by the stonework.*

OPPOSITE: *Progressive collapse of the burial chamber ceiling.*

Section through burial chamber of the Step Pyramid

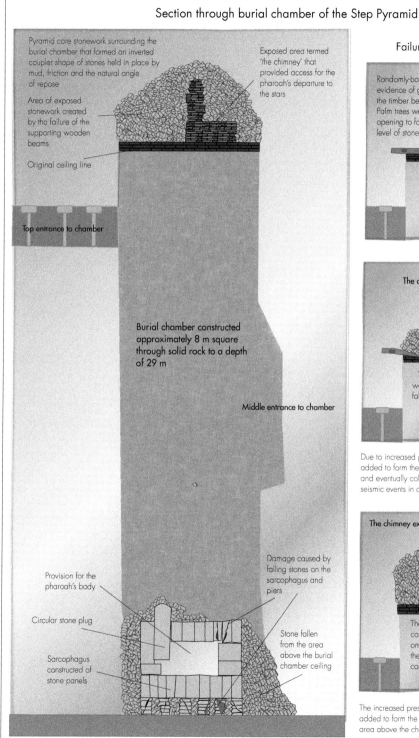

Pyramid core stonework surrounding the burial chamber that formed an inverted coupler shape of stones held in place by mud, friction and the natural angle of repose

Area of exposed stonework created by the failure of the supporting wooden beams

Original ceiling line

Top entrance to chamber

Exposed area termed 'the chimney' that provided access for the pharoah's departure to the stars

Burial chamber constructed approximately 8 m square through solid rock to a depth of 29 m

Middle entrance to chamber

Provision for the pharoah's body

Circular stone plug

Sarcophagus constructed of stone panels

Damage caused by falling stones on the sarcophagus and piers

Stone fallen from the area above the burial chamber ceiling

Failure of the wooden beams

Randomly-bonded stones in the pyramid core. Some evidence of grouting was visible above and close to the timber beams.
Palm trees were positioned above the chamber opening to form a flat ceiling and support the first level of stonework.

burial chamber

The chimney would be created as the stone blocks fell

The random coursed stones would detach from above and fall on top of the sarcophagus.

burial chamber

Due to increased pressure exerted by the additional layers added to form the Step Pyramid, the timber beams deflected and eventually collapsed. This was also accelerated by seismic events in antiquity.

The chimney expanded as the blocks continued to fall

The random coursed stones continued to detach and fall onto the sarcophagus. This was the position at the start of the contract to repair the ceiling.

burial chamber

The increased pressure exerted by the additional layers added to form the pyramid eventually caused the whole area above the chamber to collapse.

Photograph showing the amount of deflection of one of the original timber beams.

therefore chose to use the only method available to them at that time, which was timber beams. Probably, during the construction of the original mastabas 10 metres high, this would have been sufficient to hold the relatively small weight of stone needed to complete the burial chamber. However, to rely only on timber beams to carry not a single storey but three more, followed by a further three on top, was impossible. The results would not have taken long to become apparent. The time it took would depend on the seasoning of the timber and how long it took to dry out and lose its moisture, causing the problems that were about to occur.

The deflection of the beams would have started in the centre at the maximum point of bending and continued the entire length. A characteristic of timber is that when it starts to deflect, it never returns to its original position. This deflection grows incrementally and continues until eventually the beams fail, the process probably accelerated by any seismic events at the time. The stones that remain would form a small supporting arch interlocking to its natural angle of repose (the angle where the stones would self-lock).

This would be the beginning of the progressive erosion and collapse of the arch as the wedged stonework mass moved with seismic, temperature, moisture and human activity in the chamber, and stones would continue to dislodge and fall to the bottom of the shaft.

I was now more confident that a solution to stop the degradation of the chamber ceiling would be possible using our technology, but there were other solutions on the table at the Supreme Council of Antiquities that were much further down the road of acceptance. The politics of acceptance are always a mystery, given that a proposal requires approval by a large committee consisting of many members with different views and training. Not unusually, almost two years went by without any more enquiries from Intro Trading concerning movement on the project. We were then asked to give a presentation of how we could provide a complete intervention to save the burial chamber ceiling, following the very good reports and recommendations of the work we had completed saving the Temple of Hibis.

At that time in 2007, the front-runner method to repair the burial chamber ceiling was to use a complete wooden platform. This would be constructed from the base of the chamber and over the sarcophagus, rising to the defective ceiling and wedging the stone in place. It was not quite clear how this was to be achieved, but the drawings and sections I was shown were very elaborate and provided a permanent viewing platform for tourists at some later stage.

On consideration, as a close inspection of the ceiling was impossible, I asked the question: why did the timber scheme fail after it had been in the melting pot for at least the last two years? The answer was, as I thought: how do you hold up the ceiling without unlocking another and probably fatal fall of stone on the builders' heads? From the limited photographs we had taken at a distance, the ceiling was multifaceted with no straight surfaces. The use of traditional mechanical or wooden props would introduce pressure into the very dangerous structure, like a giant game of KerPlunk. The method of supporting the ceiling had to be something new and not previously considered. It had to have a large surface area and, most importantly, exert no pressure on the dangerous hanging stones, and where possible it should follow the contours of the stones.

The answer and solution came from a very unusual source. It was from the latest blast mitigation product I had patented to counter IEDs (improvised explosive devices) using the power of plain water. The product was known as Waterwall and had been developed in response to a request from the security services for rapid reaction counter-measures. We had cooperated with both British and American security services to provide a defence to strengthen existing structures against damage from high explosives and had worked successfully with its application on a number of high-risk structures in the north-east of the USA, nuclear facilities and commercial gas plants.

Following a number of explosive tests we had carried out for the British security forces, I was asked if I could provide immediate protection for installations and/or personnel against improvised explosive devices, particularly dirty bombs. Clearly this could not be our anchoring system, which takes weeks to install, and would not be suitable.

Dramatic first close-up view of the condition of the dangerous stonework burial chamber ceiling, photographed when the first scaffolding was positioned. Notice the deflect timber beam.

The stone debris and timber shoring deposited on top of the sarcophagus after the 1992 earthquake.

It was then that I thought of my inspirational hero, Sir Barnes Wallis, who invented the famous bouncing bomb used by the dam busters in the Second World War. He had worked out that the bomb had to be in contact with the dam wall to be effective; any water between the bomb and the dam wall would have a cushioning effect and absorb the explosive energy. Water, having a mass that is two-thirds that of concrete, was an ideal method of providing adequate protection. The problem was how to cause water to stand vertically when it always wants to stay horizontal. This was achieved using PVC coated wall panels, internally reinforced to create shear stress when filled with water. The product is known as Waterwall and is in operational use by the police and armed forces and others worldwide. However, where water on its own cannot be used without an additional support, there are certain applications that use the same material but inflated only with air. The use of air to lift heavy weights such as trains and aeroplanes is well established, but the usual method would not be suitable in these circumstances as the amount of movement would make it no better than using traditional props.

My first idea, in December 2009, was to use a piston-type device consisting of a hexagon shape made from reinforced drop-stitch material containing an inner bag that worked on a jack-in-a-box principle and could be inflated separately to take the shape of the irregular contours of the burial chamber ceiling. This was used in the first outline presentation to the Supreme Council and established the principle of holding up the ceiling stones. The proposal for the provision of temporary works and lifting equipment by the scaffolding firm SGB Egypt was a draft document for discussion purposes only, and was to clarify, agree and outline a proposal to gain access to the ceiling. As there was a demarcation between the operations to be carried out the project was split into two parts.

The first phase of the project would be to clear the rubble at the base of the burial chamber and survey the possibility of providing a steel scaffold to the underside of the damaged ceiling. The use of a lift or a hoist to remove the fallen stone from the top and surround of the sarcophagus was considered, but later rejected by the Supreme Council of Antiquities.

The greatest caution was needed when working in the main burial chamber to prevent stones falling upon men clearing the rubble and constructing the scaffolding. This was to be achieved by using a net or cages so that the workmen could be under cover at all times. The protection of the sarcophagus was also discussed after the rubble was removed, prior to any scaffolding being erected to the full height of the chamber.

Phase 2 of the programme would be the installation of 140 K/N (14 tonnes) of scaffolding to approximately 2 metres below the spring point of the defective ceiling, strong enough to support all the load imposed by the air jacks, workforce and equipment. We would then work on a safe-area strategy, which would allow the creation of small protected areas.

Again, after a delay of several months and more questions about the method of repairing the ceiling, we were asked to visit Cairo and meet the chairman of the Supreme Council to discuss our proposals in detail. The main issues to be examined were the method of support to be used. Would it be able to support all the weight required? Where would the cages be positioned? Would they break under load? How long could they be used in position? And many more questions regarding the suitability of the propping of the ceiling. At that meeting with the chairman, Mr Zahi Hawass, and a number of engineers, the relevant archaeologists discussed our proposals and it was agreed that a complete concept specification and drawings be prepared for final approval.

Whilst assembling all the information needed to prepare this report we hit the first major problem. I had been in contact with SGB and had had discussions with their manager, and I had requested a proposal to provide a complete scaffold from the base to the top of the chamber. Mr Hawass now said that they were no longer interested in the project. Their engineer considered it too dangerous for them, and they declined the offer to provide a scaffolding solution. This was a major setback, and an alternative had to be found without further delay.

The next problem was the air support bags that were designed and tested for a one-ton load. These would not be sufficient following Richard Swift's revised calculations and needed to be upgraded to three tons, if the project was to continue.

Mike Preece, my Waterwall production manager, was asked to establish a suitable air pillar to take the load. We were at the disadvantage of not knowing the total height of the air bags, although there were a number of drawings from a scan showing the contours of the eroded ceiling. It would be impossible to make individual air bags to support the ceiling without the fixed-data point of the proposed scaffold.

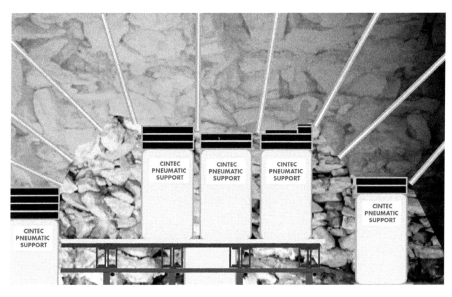

The Cintec repair concept originally presented to the Council of Antiquities.

This required some careful thought. We needed a solution that addressed all the arguments. The air bag had to be of sufficient diameter to reduce the load exerted on the precarious stone ceiling. An optimum standard height had to be established without detailed information about the height between each individual stone and the scaffold base platform.

This meant that when the contract began and the scaffold and working platforms were in place, the air bags would need to be individually shimmed into position using timber platforms, positioned beneath each bag, to place the air bags into their final positions. The bags were made to a standard size of 1.5 metres high by 1 metre diameter and would need to be inflated to 8 pounds p.s.i.

It was also necessary to change the direction of the internal reinforcement in the air bags so that rather than being positioned horizontally within the air bag (as in the first design) they were positioned vertically in layers similar to a wedding cake, so that the overall height of the air bag could be controlled to within ±1 mm. This was vital for the system to succeed. Without this internal reinforcement within the air bag, it would simply inflate like a balloon, without any dimensional control, and exert a wrongly placed pressure on the stone ceiling, resulting in inevitable collapse.

The next problem was to address the issue of the interface between the multifaceted surface area of the damaged stones and the top of the air bags.

A 1mm-thick rubber layer was glued to the top of each bag, and we acquired a good number of high-density black foam sheets varying in thickness from 50 mm to 200 mm to be cut and shaped around the individual stones when the bags were finally positioned. This would provide direct contact between the air bags and the ceiling, transferring the load from the ceiling to the scaffold platform.

Placing the air bags in their final position could only be completed when all the other elements were in place. The scaffold needed to be secure and ridged, and the temporary shimming timber supports ready and waiting in the burial chamber. The final operation to position the air bags had to be undertaken very carefully by hand before the air bag was fully inflated.

To manoeuvre the air bags into position we first inflated them to 1 p.s.i.; this was to provide their rough overall shape. Then, while they were still very pliable, they were easily positioned on top of the timber shims in the final support location, and when in position, we inflated them to 8 p.s.i. to achieve the 3 tons loading that was required. To assist the deployment of the bags, we incorporated an air control mechanism, so that all the bags were interconnected. Each bag could be inflated individually or as a group, should the need arise.

At the planning stage, without any direct access to the ceiling, it was impossible to judge the ceiling's condition, or the individual strength or weakness of each group of stones. It was initially assumed, that the air bags should be positioned equally throughout the burial chamber. However, when we finally had access to the ceiling, it was Dennis who made the final decision, looking up from a distance of 300mm to the ceiling, on where he would position the air supports to give maximum protection to the technicians. In the event, there turned out to be three nearly flat stones in the very centre of the chamber ceiling, and we put three almost contiguous air bags beneath these because this was the position in most danger of collapse.

ABOVE: *An air support bag prior to testing at Cintec's office.*

RIGHT: *The first air bag in position in the burial chamber.*

ABOVE LEFT: *Control box used to regulate and control the air pressure in all the air bags.*

ABOVE RIGHT: *Three air bags in position at the centre of the ceiling.*

LEFT: *A dramatic photograph showing how dangerous the hanging stones were during the positioning of the air bags. The first jigsaw view from the bottom of the chamber was only a 2D view not a 3D close-up view providing another dimension to the blocks.*

OPPOSITE: *Step Pyramid of Djoser at Saqqara under renovation.*

The then preferred method of permanently securing the dangerous ceiling was using our socked anchors. The installation method was discussed in great depth. These were to be at a metre-grid pattern throughout the burial chamber. These anchors would be installed roughly at 90° to the normal of the extremely rough-shaped arch and were designed to lock and transfer the load into solid material at a higher level. The anchors were made from 20-mm stainless steel-threaded bars installed in a 52-mm dry diamond-drilled hole incorporated in a fabric sock to contain the designed grout in the desired position, and also to bridge any voids or openings formed by the collapsed stonework. The anchor lengths varied between 3 to 4.5 m depending on their location, and together with an examination of the individual drilled core log provided an instant report on the condition of the inner stonework above the ceiling.

The secondary method of securing the ceiling was to strengthen an area greater than the size of the burial chamber. If the burial chamber was 8 m x 8 m, we strengthened an area 12 m x 12 m or more using the system. It would provide a large plug on top of the chamber stopping the fall of masonry into the chamber.

Another important consideration when executing the anchor injection in this procedure was the problem of pumping the grout into an anchor that was to be installed at an angle above the horizontal. This means that the technician is pumping grout into an anchor above his head and the grout has to flow uphill to fill the assembly completely. In practice, the pressure injection system is pumped via an air compressor into a pressure pot and then through an injection pipe and nozzle into the anchor. The grout delivery pipe in the anchor goes from the front plate of the anchor to the plate at its far end, well within the damaged structure. As the grout is pumped into the assembly, which is in a blind hole, a problem could arise, since, as the grout is pumped in, any trapped air would gather and be compressed at the innermost end of the anchor. This would eventually match the pressure of the air compressor, and the result would be that the anchor would not be completely filled.

To overcome this problem, we attached a Ventura device to the pressure pot, which in turn was connected to a thin tube that paralleled the grout delivery tube and terminated at the far inner end of the anchor. This provided a vacuum in the assembly emanating from the innermost end of the anchor, making an area of low pressure that obviated the problem of trapped backpressure in the intervention.

At the beginning of the procurement process we had signed an agreement with the Green Bay Film Company from Cardiff who were engaged by *National Geographic* to film the restoration with a project title of 'Saving Egypt's Oldest Pyramid'. A tester film had already been produced on the small repairs undertaken at the Red or North Pyramid at Dahshur, and the company were very anxious, spurred on by *National Geographic*, to start filming.

The vagaries of contracting for such an iconic monument with all the participants wanting the project to be completed successfully, but without haste on the client's part, proved very alien to the media company. They wanted the work shot and in the can and wrapped up as soon as possible.

On 28 July 2009, the concept proposals were given to the main contractor's consultant, Dr Ashraf El Zanaty, for approval of the proposed sequence of the work. He in turn arranged with Dr Hassan Famy, the project consultant for the Supreme Council for the formal acceptance of the scheme.

Whilst these proposals were being considered by the Supreme Council of Antiquities for technical approval, we also had to investigate the bottom end of the chamber and the myriad problems that could arise in providing a stable platform to erect sturdy scaffolding.

The first problem was to remove all the debris of fallen stone and timber that was shaken loose in the 1992 earthquake. There must have been 150 tons of debris covering the sarcophagus at the base of the chamber. It was vital that this be removed so that we could see the condition of the sarcophagus and its surrounding area and provide a solid foundation for an independent scaffold. We had proposed removing the debris from the top of the chamber using a winch located and controlled from the main top entrance. Curiously, this was not the method adopted by the Supreme Council, as they believed it could damage the corridor floor at the main entrance owing to the weight of the heavy equipment. This was not what I had proposed, and there must have been some misunderstanding about the proposal or its translation. Having spent eighteen years of my life in the navy serving in mine sweepers, I was very conversant with the use of winches, which are the sole means of moving large and heavy weights on board a ship. I had proposed that the heavy winch be situated outside the pyramid but the lifting wires would be directed into the pyramid with snatch blocks to direct the movement of the wires from the winch to the top of the chamber, where only a single roller needed to be positioned to lift the individual stones from the base of the chamber. Needless to say, the exact opposite was carried out, and the stones and timber were removed via the lower tunnels to ground level outside the pyramid. This was slow and backbreaking work that caused a long delay in the programmed scheme's acceptance.

Once the sarcophagus had been exposed, a further setback was discovered. The piers or supports holding it up were badly damaged by the fall of rock in the 1992 earthquake, 150 tons descending 29 metres onto the sarcophagus, which on its own must already have imposed 80 tons of self-weight on the piers. This load transferred straight through the sarcophagus to the piers, causing them to twist, buckle

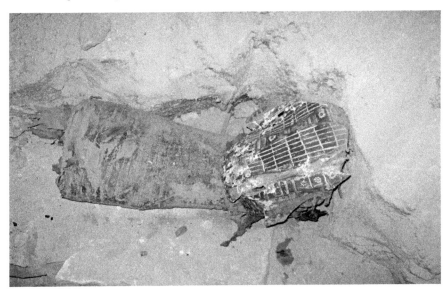

Unrelated mummies discovered in the chamber but much later in origin.

*Condition of the loose
stonework.*

and crumble. A further complication, as if we needed any more, was the problem of what might lie beneath the sarcophagus. With the myriad known tunnels in the complex, could there be an unknown tunnel directly beneath the area we wanted to work in, caused by a robber? This all had to be investigated and the sarcophagus piers reinstated by the main contractor.

At the time of the excavation, at the sides of the sarcophagus two mummies were found and examined by the Supreme Council of Antiquities. I understood later that unknown persons placed these mummies at a later date, and they were not contemporary with the Pharaoh Djoser.

The structural integrity of the piers was restored using conventional methods and then, layer by layer, sandbags were placed around and alongside the sarcophagus to protect it and consolidate the base area to receive the independent scaffolding.

While the preparatory work at the base of the chamber was being undertaken by the main contractor we had time to find a new scaffolding subcontractor who was able to undertake this very delicate work.

The consultants for the Supreme Council were intrigued with the proposal to hold up the defective ceiling with air, but they were also hands-on practical engineers who wanted to see at first hand the product and its ability to perform as advertised.

Later, in November 2010, a party of consultants led by Dr M. Ghamrawy and Dr Ashraf El Zanaty arrived at our head office in Newport to watch the compression tests on each bag and to sign that each one of them was fit for purpose. Normally, this would be done in our factory, but to help our guests to acclimatise to the 50 mm of snow that had fallen overnight and to observe the individual tests, we arranged to have the test rig positioned outside our main office boardroom French window. These all proved successful and the air bags and test certificate were duly signed.

Upon their return to Cairo, the Supreme Council had requested a visit from the current UNESCO structural engineering consultant for the Step Pyramid to review the proposed strengthening work necessary on the distressed ceiling. This was carried out by Professor G. Croci from Italy, who inspected the ceiling and confirmed that the method we had proposed to finally secure it was the best method of undertaking the work.

The next major issue was the scaffolding. This was needed to provide a working platform from the base of the burial chamber to just below the spring points of the defective arched stonework. We needed a company with an international reputation which would be able to supply a scaffolding system that was well proven and tested, and most importantly that we could trust.

The company chosen was Acrow Misr an international company based in Egypt. Working with Acrow, we were able to approve a suitable design that encompassed all our requirements for both structural strength and usability. When this was finally agreed, we submitted the final drawings and specifications to the Council of Antiquities for their approval. This was later confirmed and the project proceeded.

Normally, work of an indeterminate nature such as this is undertaken on cost-plus basis because of the very nature of the work: the unknown quantities involved that cannot be estimated before the scaffold is in position; also, the position and location of temporary air supports are a guess at best and therefore are normally under a schedule of rates contract.

This was not to be, and the client would allow only a fixed price to complete phase 1 of the project. After a great deal of time carefully assessing what problems we would encounter and what conditions we might expect, a fixed price contract was submitted to Intro Trading for onward transmission to the main contractor and the authorities and the contract was awarded in December 2010 for better or worse.

The actual start date would be almost six months later when the scaffold was erected and in position. All the necessary materials, including stainless steel, fabric sock and grout manufactured to match the parent material, were acquired and delivered to Cintec's factory in Newport. The special drilling diamond drills, extensions, cutting head and a whole host of various items of equipment were ordered in the United Kingdom ready for packaging and transport to Cairo.

The race had started.

We had by this time some of the first photographs of the damaged ceiling area taken close up from the freshly erected working platform. These photographs were graphic and showed all the stonework in 3D rather than the 2D which we had first viewed from the base of the burial chamber. Not only had the stones developed a third dimension but they had also grown considerably in size and shape.

At the very centre of one of the photographs there was an opening like a chimney flue, which did not seem to have any visible support and looked particularly dangerous. This was known by all, as the chimney. It was later believed to be the access route of the pharaoh's Ba when it ascended or returned from the stars from his earthly body in the sarcophagus. It was of strange appearance for an access point for a god. Perhaps a decorated opening to the vertical flue had disappeared when the original ceiling had collapsed; there was no indication on what remained that it was special or significant.

ABOVE: *The most dangerous 'chimney shaft'.*
RIGHT: *Commencement of deep pointing the loose stonework.*
BELOW: *Smaller bespoke air bags were used at the perimeter of the chamber.*

Our first thoughts were that it was an access shaft to another chamber over the main chamber. When we were able to gain access to the shaft it seemed to terminate about 3.5 metres above the then ceiling position. A naked flame would flicker in the air stream indicating that there was another source of air a metre up the shaft. However, we could not get any higher up as it seemed to corbel over at the top. This coincided with the top of the first mastaba position. Also, when we were able to drill vertically 4 metres above the ceiling, there was no sign of any other chamber, and the drilling cores did not indicate any large opening.

With great trepidation, our first working day on site began. All the scaffolding had been positioned, some altered at the last minute to stop the scaffolding from swaying under the weight of everyone who was able to be on site that day, including the film crew watching from the access platform. It was not possible to position the air bags in the position we had originally intended because of the profile of the ceiling when it was examined close to its surface. Luckily, there were three reasonably flat stones in the very centre of the ceiling and we decided that it would be best to support these slabs first and as quickly as possible to provide some measure of support to the ceiling. Dennis Lee and Mike Preece were first in the firing line to secure this vital support. Dennis told me later that working in a confined space with 62 metres of hanging stone delicately balanced over his head while he cut the ridged structural foam to size with a hand saw and wedged it in place was absolutely terrifying.

To place the air bag in its exact position under the stone he could not inflate it to more than 1 p.s.i. This was because he needed a semi-ridged air bag that could be easily manoeuvred and positioned in the precise place. He knew that it would not have been able to support the weight of the stonework above him should there have been a fall of stone. The air bag had to be inflated to 8 p.s.i. to take the three-ton designed load. Luckily, we were able to fix the first three main air bags on the three central flat stones within hours, and a great sigh of relief was heard throughout the burial chamber.

It was very obvious then that there were not enough to secure the ceiling. The eleven original air bags were 1 metre in diameter and 1.5 metres high. They would not fit at the perimeter of the chamber because they would be too high. That was a problem that did not need to concern me at the moment, and the team continued over the next few days positioning and fixing all the eleven original air bags into place.

When this was achieved, everyone was relieved. At least we had succeeded in our first objective and now we could start to plan the next move.

The next problem was a case of measuring the additional air bags needed around the chamber perimeter. This proved to be another sixteen air bags of the same diameter, but only 1 metre high. This again had to be approved by the Supreme Council as an addition to the contract. Once the approval was obtained, the bags were fabricated in the factory in Wales and then sent by air via the Egyptian Customs to the Step Pyramid.

On careful examination of the state of the stonework at close quarters, the haphazard jumble of stones, particularly in the areas where the most movement had taken place, confirmed our impressions that any attempt to drill the stones would result in total failure. Whilst the method chosen was dry diamond drilling,

any attempt to do this would result in the stones rotating with the drill head, again causing a ceiling collapse. The stones had to be fixed in place without any movement when drilled. From the evidence at long range, we had nominally priced for 75 mm of pointing, using a sympathetic lime grout mixture. In reality, this proved totally inadequate. The depth of pointing in some instances, was up to 400 mm. Cavities found by stretching an entire arm into the gap were found on a regular basis.

As always, in Cairo each step of the way has to be approved, normally with empirical tests to prove that it is fit for purpose. The next challenge was to prove that the lime grout we had recommended to point the stonework was acceptable to the client. We had previously sent data sheets to the authorities giving the strengthened values of the grout that we proposed to use. We had considered using the local available lime grout mixtures, but its quality did not compare with the quality available from our normal supplier, Conserv Natural Hydraulic Lime, grout code NH5. This we had used with great success on other historic buildings. A compromise was achieved, using locally obtained aggregate, mixed with Conserv's lime. The test results from Cairo University proved excellent, particularly, the compressive strength and elastic moduli needed on such heavy stonework with wide gaps between them.

The average results after 14 days were;

Compressive strength	43.4 kg/m^2
Tensile strength	5.7 kg/m^2
Bond strength	4.4 kg/m^2

This test was repeated and the compressive strength increased to 57 kg/m^2.

Thus, the painstaking evolution began, of carefully removing the 4,700-year-old mud from between the individual stones, using a water-filled fine nozzle spray, attached to a small indoor plant-watering flask, carefully removing the mud and allowing the stone to dry. Individual stones dried almost instantaneously in the carefully controlled environment of the burial chamber, which was kept at a temperature of between 20° and 24° and at a constant 35 per cent humidity.

These conditions were not easy. With all the air bags fitted, the space in the chamber was considerably reduced by their large diameter; the space needed to carry the load in those early days gave very little room for manoeuvre, especially considering the additional scaffolding required for an extra level placed on top of the main working platform for higher access.

The constant coming and going of officials and interested parties to observe the burial chamber close up, many of them for the first time in their careers, was also a hindrance. At one time, there were so many people in the chamber that I thought that I was at a reception and someone was going to hand me a sherry.

National Geographic *filming the work as it proceeds.*

Our Public Relations consultants, Rob Petersen's, were also working flat out to get the maximum advantages from the great interest the project attracted. Once the local press had published the details of our work the multinationals and other foreign press, all joined the melee. I remember the *Independent* claiming, 'If you want a Pyramid repaired go to Wales.' BBC Wales were very kind to us with local TV coverage, and I was asked to do two radio interviews, one at the Llandaff studio and one live from the pyramid at Saqqara.

The author inspecting the air bags.

Probably, the most amusing interview, again on the radio, but for BBC America, was with an American lady interviewer. She had a pronounced American accent, and having talked through the project in detail, finally asked, 'What did you think of the curse of the pharaoh, working in the burial chamber'? I replied that the only curse would be if one of the stones fell on our heads.

In the meantime, the initial tests on the anchor were arranged with Cairo University to test three anchors embedded in the stones that had fallen from the chamber and could be classified as typical parent material in the chamber ceiling.

The three 20-mm anchors were installed in 52-mm diameter diamond-drilled holes with an embedded depth of 155 mm on one anchor and 180 mm on the other two samples. These were tested to an ultimate load of 20220N and 24850N respectively, well within the design considerations.

The pointing phase was due to continue from September 2011 until almost Christmas of 2012. The work was exacting and uncomfortable, and carried out in very cramped and dangerous conditions. Dennis Lee and his team did a remarkable job.

Questioning how the anchors would work when fixed in the burial chamber ceiling that was full of varying-size voids was the next hurdle to overcome. The consultants were very happy with the results of the tests completed at Cairo University, but how would the anchors behave installed in a wall full of voids?

These were bread-and-butter everyday questions I am asked continually worldwide. The system was designed for this very problem. The fabric sock that encapsulates the anchor body retains all the injected grout within the assembly and it is able to pass over voids without the problems. Without the fabric sock a single anchor body would have spilled the grout everywhere and not functioned as an anchor.

To prove this, a further test was set up outside the pyramid. This test illustrated how the system worked. Two 52-mm diamond-drilled holes were made through two stone blocks from the pyramid, similar to cavity-wall construction in a modern house. A large gap was created between the blocks, and two anchors of the diameter to be used on the contract were made to lengths that suited the test blocks. The anchors were then inflated in the normal manner and allowed to cure for seven

Diamond drilling test in stone blocks at the entrance to the pyramid before setting test anchors.

days. The proof load required was 17K/N (1.7 tonnes) per anchor and this was easily achieved producing an acceptable displacement of 0.82 mm on the front block and 0.55 on the rear block. The ultimate load achieved was 79.95 K/N. (8 tonnes). This was the final proof required by the Supreme Council to give permission for the first test hole to be drilled inside the burial chamber of the pyramid.

This ratcheted up the tension and pressure on all the construction team, as not only was it vital for the continuation of the project but also that we moved to the next trigger point for payment on the contract.

Under the spotlight, with the *National Geographic* film crew filming for global coverage every move, gesture and facial expression the team made did not calm the nerves.

The day and the time had been set, and a position just inside the main burial chamber a metre above the working scaffold platform had been chosen by the consultants. From this vantage point, they could easily monitor the trial drilling, the view only impeded by the scaffolding and not the air bags.

A vibration monitor had been installed close to the drilling equipment to monitor the vibrations from the dry 52-mm diamond drill core to be used to create the first hole in the 4,700-year-old monument. The tension was electrifying, not helped when an ambulance arrived, courtesy of the client, just in case we were trapped by a fall of rock initiated by the drilling process, and needed to be whisked off to hospital. With all the people who attended the test, I just wondered, if the worst came to the worst, who they would drag out and who they would leave.

Once the rigs were set up and the dust extractors put in place, the power was switched on. Dennis was taking no chances and was working the drill himself, holding it gently, in the temporary jig we use to start the core in the correct position.

The drill started, and the 52-mm core bit rotated slowly with its rhythmic chatter associated with dry drilling. The drill was stopped momentarily and the jig was removed. The drill returned to the hole and started again. So far, so good. The drill continued, producing the inevitable cloud of dust that escaped the extraction vacuum. All was going well and the drill had penetrated 150 mm into the limestone vertical burial chamber wall, when a sharp noise was heard and the drill bucked slightly. Dr Hassan, the project consultant, requested Dennis to stop drilling whilst he checked the vibration-measuring meter. You could have heard a pin drop. The meter indicated that there was a small vibration, but it was not over the limit he had set as acceptable.

Dennis explained to him that this was not unusual, as sometimes small hardened fragments are found in the limestone and the drill would pass through them in a second or two.

With permission, he restarted the drilling and drilled continuously, and without any further mishap, until the core bit had to be extended to complete the full length of the hole. I am sure that if the pharaoh had been present he would have also wanted the work to continue to restore his earthly home.

We were in business.

The contract then started in earnest with the drilling method approved and all the material at hand. Unfortunately, what was immediately obvious, and we had known this for some time, was that the original grid pattern proposed before we had inspected the stonework would not be sufficient to secure the ceiling. During

The completed central section of the ceiling after the removal of the air bag supports.

ABOVE: A photograph of the only stone that moved and rotated when the work was in progress. This cleared the burial chamber faster than Usain Bolt's 100-metre run!

ABOVE AND RIGHT: Installation of anchors at the perimeter of the chamber.

the pointing element of the work it was clear that a regular anchor pattern could not obtain the permanent result that we were hoping to achieve. Again, the three-dimensional aspect of the surface and the haphazard profile of the stones proved that each stone had to examined, and anchors would be needed as and when required.

This was agreed by the consultants and work commenced to consolidate the defective ceiling.

The safest method was to start at the centre, on the few flat stones that were exposed next to the three main air bags. The first 3.5-metre anchor was drilled vertically, straight above, at 90°. We had hopes of finding another chamber, but the core samples proved negative and the cores indicated that the stones immediately above the ceiling were in semi-regular courses with each stone not exceeding 300 to 400 mm in size. The other safety method we adopted was to drill and install only one anchor at a time, which in practice meant that we could only install one anchor per day.

It was a great relief when we had installed a group of central anchors in the most vulnerable section of the ceiling, which meant that if there was any structural movement taking place in the unstrengthened stonework, this group of anchors would provide some limited protection for the team.

From then on, it was a matter of working around the inner perimeter of the main large bags, installing anchors at lengths and in directions determined by the position of the stones. The position of the anchors was recorded from base lines set up on the inside of the burial chamber walls by giving each anchor a number and a coordinated position relative to the base lines.

Work progressed slowly but surely, and the anchors were installed between the supporting air bags until all the central area was completed. The next act of faith was to remove each air bag in turn, pointing the exposed area and installing the necessary anchors. This provided more working space and the conditions improved immensely. The difficulty now was at the perimeter of the chamber, which still had 4,700 timbers fossilized in position, and it was the chimney area that caused most concern.

The chimney area and particularly the stones at the opening proved a different matter. Because the stonework at the breakdown of the opening had been subjected to greater movement and because of the progressive failure of the bond between the stones as they collapsed inwardly down the shaft, they were unsupported on one side, and sometimes on two.

A heart-stopping moment came whilst Dennis was removing mud from a projecting stone at the entrance to the chimney. The London Olympic Games were in full flow at the time, and I recall that the men's 100 metres event was in progress. This dominant stone was some 300mm high, 250mm wide and 1500mm long, stretching across the chimney opening. The work to remove the accumulated coating of mud it had gathered during the last 4,700 years, was underway, gently, spraying a water mist across its surface, when, without any warning, it rotated and fell out at one end towards the scaffold platform and Dennis. The resultant dash to evacuate the chamber was, in my opinion, a quicker time than the race in London. Gingerly, the team returned and propped the stone, and work returned to normal, or so we believed.

Events totally beyond our control were about to happen. Although, by, then we had installed all the scheduled number of metres of anchors for phase 1 of the project and provided all the estimates to continue to the second phase, work abruptly stopped because of the rebellion and the civil unrest with a change of government on 18 March 2013.

A year later, whilst the melting pot of Egypt's political scene boiled over and the position regarding the completion of our work was unclear. I read in the Royal Chartered Institute of Buildings magazine that an inaugural international competition was being launched in June 2014. There were a number of categories for the Outstanding Achievement awards for project management and innovation, encompassing fields as diverse as procurement, conservation, design, people management, facilities, and health and safety management.

I was very interested in the conservation category. This aimed to discover the unsung heroes in the project teams who have made an

ABOVE: *The author receiving the International Award for Conservation from the Chartered Institute of Builders for work related to the project.* OPPOSITE: *The Step Pyramid complex of Djoser showing the chapels on the west side of the Heb-Sed Court in the foreground.*

outstanding contribution and have changed things for the better in the world of construction. Having prepared a submission on the structural repair to save the burial chamber of the Step Pyramid, I only needed an endorsement from a respected member of the Institute for the completion of my application. I asked John Edwards, a chartered surveyor who was well-respected in the field of restoration. I had worked with him in the past on ancient monuments in Cardiff. He agreed and provided the necessary endorsement to my submission.

Several months later, out of the blue and much to my surprise, I received an email stating that I had won the inaugural International Award for Conservation.

I was the only United Kingdom recipient of any award, the other category winners coming from Canada, Ghana and China. The presentation was held in November 2014 in London.

BIBLIOGRAPHY

Admiralty Great Britain, *Admiralty Manual of Navigation*, 2. (London: HMSO, 1964).

Arnold, Dieter, *The Encyclopaedia of Ancient Egyptian Architecture* (London: I.B. Tauris, 2003).

Ashurst, J. and Ashurst, N., *Practical Building Conservation — Volume 1: Stone Masonry* (Aldershot: Gower Publishing, 1988).

Ashurst, J. and Ashurst, N., *Practical Building Conservation — Volume 2: Brick Terracotta and Earth* (Aldershot: Gower Publishing, 1988).

Ashurst, John, *Conservation of Ruins* (London and New York: Routledge, 2007).

Baer, N. S., Livingston, R. A. and Fry, S., *Conservation of Historic Brick Structure: Case Studies and Reports of Research* (Dorset: Donhead, 1998).

Barke, James, *Egyptian Antiquities in the Nile Valley* (London: Methuen, 1932).

Barocas, C., *Monuments of Civilization: Egypt* (London: The Reader's Digest Association Ltd, 1974).

Beckmann, Poal, *Structural Aspects of Building Conservation* (New York: McGraw-Hill, 1994).

Breasted, J. H., *A History of Egypt From the Earliest Times to the Persian Conquest* (Charles Scribner & Sons, 1916).

Brereton, Christopher, *The Repair of Historic Buildings: Advice on Principles and Methods* (London: English Heritage, 1991).

Cain, J. A. and Hulse, R., *Structural Mechanics* (London: Palgrave Macmillan, 1990).

Cartwright, K. and Findlay, W., *Decay of Timber and its Prevention* (London: HMSO, 1946).

Colston Research Society, *Cosmic Radiation*, Colston Papers based on a Symposium promoted by the Colston Research Society and the University of Bristol in September 1948 (London: Butterworth Scientific Publications, 1949).

Croci, Giorgio, *The Conservation and Structural Restoration of Architectural Heritage* (Southampton: WIT Press, 1998).

D'Ayala, D. and Fodde, E., *Structural Analysis of Historic Construction*, 1 and 2 (Florida: CRC Press, 2008).

Den Hartog, J. P., *Strength of Materials* (New York: Dover Publications, 1961).

Dodson, Aidan, *The Pyramids of Ancient Egypt* (London: New Holland Publishers, 2003).

Draycott, Trevor, *Structural Elements Design Manual* (London: Routledge, 1990).

Edwards, A. A. B., *A Thousand Miles up the Nile* (London: Darf Publishers, 1993).

Edwards, I. E. S., *The Pyramids of Egypt* (London: Ebury Press, 1947).

Feilden, Bernard M., *Between Two Earthquakes: Cultural Property in a Seismic Zone* (Rome; Marina del Ray, CA: ICCROM; Getty Conservation Institute, 1987).

Feilden, Bernard M., *Conservation of Historic Buildings* (Oxford: Butterworth-Heinemann, 1994).

Gartner, R., *Statically-Indeterminate Structures* (London: Concrete Publications Ltd, 1958).

Geddes and Grosset, *Ancient Egypt: Myth and History* (New Lanark: Gresham Publishing Company, 1997).

Geeson, A. G., *Building Science Materials*, 4 (London: English Universities Press, 1960).

Hawass, Zahi, *The Treasures of the Pyramids* (Cairo: The American University in Cairo Press, 2003).

Hendry, A. W., *Structural Masonry* (1987; 2nd edn, London: Palgrave Macmillan, 1998).

Hilson, Barry, *Basic Structural Behaviour: Understanding Structures from Models* (London: Thomas Telford Services Ltd, 1993).

Hollister, G. S. (ed.), *Developments in Stress Analysis*, 1 (n.p., 1997).

Jackson, K. and Stamp, J., *Pyramid: How and Why it Was Built* (London: BBC Books, 2002).

Jackson, K. and Stamp, J., *Pyramid: Beyond Imagination: Inside the Great Pyramid at Giza* (London: BBC Books, 2002).

James, T. G. H., *Egypt: The Living Past* (London: British Museum Press, 1992).

James, T. G. H., *Egypt Revealed: Artist-travellers in an Antique Land* (London: The Folio Society, 1997).

Jordan, Paul, *Egypt: The Black Land* (London: Phaidon 1976).

King, L. W. and Hall, H. R., *History of Egypt, Chaldea, Syria, Babylonia, and Assyria in the Light of Recent Discovery* (New York: Grolier Society, 1906).

Landels, J. G., *Engineering in the Ancient World* (Oakland: University of California Press, 2000).

Lehner, Mark, *The Complete Pyramids* (London: Thames and Hudson, 1997).

Lizzi, Fernando, *The Static Restoration of Monuments* (Genoa: Sagep, 1982).

Lourenço, P. B., and Roca, P., *Historical Constructions* (Guimarães: University of Minho, 2001).

Maspero, G., *Egyptian Archaeology* (London: H. Grevel & Co., 1889).

Maspero, G., *New light on Ancient Egypt* (London: T. Fisher & Unwin, 1908).

Massimo, Mariana, *Trattato Sul Consolidamento E Restauro: Degli Edifici in Muratura*, 1 (Rome: DEI, 2006).

Massimo, Mariana, *Trattato Sul Consolidamento E Restauro: Degli Edifici in Muratura*, 2 (Rome: DEI, 2012).

Mendelssohn, Kurt, *The Riddle of the Pyramids* (London: Thames & Hudson, 1974).

Mitchell, Eleanor, *Emergency Repairs for Historic Buildings* (London: English Heritage, 1988).

Mitchell, George, *Building Construction* (1902; 13th edn, London: B. T. Batsford, 1943).

Mitchell, George, *Mitchell's Advanced Building Construction: The Structure* (London: B. T. Batsford, 1963).

Morley, Arthur, *The Theory of Structures* (London: Longmans, 1961).

Neret, Gilles, *Description De L'Egypte* (Cologne: Taschen, 1840).

Oakes, Lorna, *Sacred Sites of Ancient Egypt* (London: Hermes House, 2006).

Parkinson, G., Shaw, G., Beck, J. K. and Knowles, D., *Appraisal and Repair of Masonry* (London: Thomas Telford Services Ltd, 1996).

Parry, Dick, *Engineering the Ancient World* (Stroud: The History Press, 2005).

Petrie, W. M. Flinders, *A History of Egypt, From the Earliest Kings to the XVIth Dynasty* (London: Methuen, 1903).

Petrie, W. M. Flinders, *Seventy Years in Archaeology* (London: Sampson Low, Maston & Co., 1931).

Petrie, W. M. Flinders, *The Pyramids and Temples of Gizeh* (1883), Histories and Mysteries of Man, with an update by Zahi Hawass (London: Leadenhall Press, 1990).

Petroski, Henry, *Invention by Design: How Engineers Get from Thought to Thing* (Cambridge and London: Harvard University Press, 1996).

Plommer, H., 'Ancient and Classical Architecture', *Simpson's History of Architectural Developments*, 1 (Harlow: Longman, 1956).

Purchase, William R., *Practical Masonry* (London: Crosby, Lockwood and Son, 1903).

Quadling, D. A. and Ramsay, A. R. D., *Introduction to Advanced Mechanics* (London: G. Bell & Sons, 1964).

Reynolds, T. J. and Kent, L. E., *Introduction to Structural Mechanics for Building and Architectural Students* (Ammanford: The English Universities Press, 1961).

Robson, Patrick, *Structural Repair of Traditional Buildings* (Dorset: Donhead Publishing, 1999).

Roca, P. and Molms, C., *Arch Bridges* IV. *Advances in Assessment Structural design and Construction* (Barcelona: CIMNE, 2004).

Salvadori, Mario, *The Art of Construction: Projects and Principles for Beginning Engineers & Architects* (Chicago: Chicago Review Press, 1990).

Shaw, Ian and Nicholson, Paul, *British Museum Dictionary of Ancient Egypt* (London: British Museum Press, 1995).

Smyth, Piazzi, *Our Inheritance in the Great Pyramid* (London: Daldy, Isbister & Co., 1877).

Stewart, C., 'Early Byzantine & Romanesque Architecture', *Simpson's History of Architectural Developments*, 2 (London: William Clowes Ltd, 1956).

Stewart, Desmond, *The Pyramids and Sphinx* (New York: Newsweek, 1979).

Strengthening and Stabilisation of Concrete and Masonry Structures (International Concrete Repair Institute, 2003).

Structural Conservation of Masonry, conference proceedings, International Technical Conference, Athens, 1989 (Rome: ICCROM, 1990).

Talbot-Kelly, R., *Egypt Painted and Described* (London: Carl Hentschel Ltd, 1903).

Tompkins, Peter, *Secrets of the Great Pyramid* (London: Penguin Books, 1973).

Tyldesley, Joyce, *Egypt: How a Lost Civilization was Rediscovered* (London: BBC Books, 2005).

Verner, Miroslav, *The Pyramids: Their Archaeology and History* (London: Atlantic Books, 1997).

Ward, John, *Pyramids and Progress* (London: Eyre and Spottiswoode, 1900).

Weaver, M. and Matero, Frank G., *Conserving Buildings: A Manual of Techniques and Materials* (New Jersey: John Wiley & Sons, 1993).

Wilkinson, Toby, *The Rise and Fall of Ancient Egypt* (London: Bloomsbury, 2010).

Williams, Clement C., *The Design of Masonry Structures and Foundations* (New York: McGraw-Hill, 1922).

INDEX